GACE
024
025

Science
Teacher Certification Exam

M000117569

By: Sharon Wynne, M.S.
Southern Connecticut State University

"And, while there's no reason yet to panic, I think it's only prudent that we make preparations to panic."

XAMonline, INC.
Boston

Copyright © 2008 XAMonline, Inc.
All rights reserved. No part of the material protected by this copyright notice may be reproduced or utilized in any form or by any means, electronic or mechanical, including photocopying, recording or by any information storage and retrievable system, without written permission from the copyright holder.

To obtain permission(s) to use the material from this work for any purpose including workshops or seminars, please submit a written request to:

XAMonline, Inc.
21 Orient Ave.
Melrose, MA 02176
Toll Free 1-800-509-4128
Email: info@xamonline.com
Web www.xamonline.com
Fax: 1-781-662-9268

Library of Congress Cataloging-in-Publication Data

Wynne, Sharon A.
    GACE Science 024/025: Teacher Certification / Sharon A. Wynne. -2nd ed.
    ISBN 978-1-58197-529-1
    1. GACE Science 024/025.   2. Study Guides.   3. GACE
    4. Teachers' Certification & Licensure.   5. Careers

Disclaimer:
The opinions expressed in this publication are the sole works of XAMonline and were created independently from the National Education Association, Educational Testing Service, or any State Department of Education, National Evaluation Systems or other testing affiliates.

Between the time of publication and printing, state specific standards as well as testing formats and website information may change that is not included in part or in whole within this product. Sample test questions are developed by XAMonline and reflect similar content as on real tests; however, they are not former tests. XAMonline assembles content that aligns with state standards but makes no claims nor guarantees of teacher candidates making a passing score. Numerical scores are determined by testing companies such as NES or ETS and then are compared with individual state standards. A passing score varies from state to state.

Printed in the United States of America

GACE: Science 024/025
ISBN: 978-1-58197-584-0

Table of Contents

Great Study and Testing Tips!

What to study in order to prepare for the subject assessments is the focus of this study guide but equally important is *how* you study.

You can increase your chances of truly mastering the information by taking some simple, but effective steps.

Study Tips:

1. <u>Some foods aid the learning process</u>. Foods such as milk, nuts, seeds, rice, and oats help your study efforts by releasing natural memory enhancers called CCKs (*cholecystokinin*) composed of *tryptophan*, *choline*, and *phenylalanine*. All of these chemicals enhance the neurotransmitters associated with memory. Before studying, try a light, protein-rich meal of eggs, turkey, and fish. All of these foods release the memory enhancing chemicals. The better the connections, the more you comprehend.

Likewise, before you take a test, stick to a light snack of energy boosting and relaxing foods. A glass of milk, a piece of fruit, or some peanuts all release various memory-boosting chemicals and help you to relax and focus on the subject at hand.

2. <u>Learn to take great notes</u>. A by-product of our modern culture is that we have grown accustomed to getting our information in short doses (i.e. TV news sound bites or USA Today style newspaper articles.)

Consequently, we've subconsciously trained ourselves to assimilate information better in <u>neat little packages</u>. If your notes are scrawled all over the paper, it fragments the flow of the information. Strive for clarity. Newspapers use a standard format to achieve clarity. Your notes can be much clearer through use of proper formatting. A very effective format is called the *<u>"Cornell Method."</u>*

> Take a sheet of loose-leaf lined notebook paper and draw a line all the way down the paper about 1-2" from the left-hand edge.

> Draw another line across the width of the paper about 1-2" up from the bottom. Repeat this process on the reverse side of the page.

Look at the highly effective result. You have ample room for notes, a left hand margin for special emphasis items or inserting supplementary data from the textbook, a large area at the bottom for a brief summary, and a little rectangular space for just about anything you want.

3. <u>Get the concept, then the details</u>. Too often we focus on the details and don't gather an understanding of the concept. However, if you simply memorize only dates, places, or names, you may well miss the whole point of the subject.

A key way to understand things is to put them in your own words. If you are working from a textbook, automatically summarize each paragraph in your mind. If you are outlining text, don't simply copy the author's words.

Rephrase them in your own words. You remember your own thoughts and words much better than someone else's, and subconsciously tend to associate the important details to the core concepts.

4. <u>Ask Why?</u> Pull apart written material paragraph by paragraph and don't forget the captions under the illustrations.

Example: If the heading is "Stream Erosion", flip it around to read "Why do streams erode?" Then answer the questions.

If you train your mind to think in a series of questions and answers, not only will you learn more, but it also helps to lessen the test anxiety because you are used to answering questions.

5. <u>Read for reinforcement and future needs</u>. Even if you only have 10 minutes, put your notes or a book in your hand. Your mind is similar to a computer; you have to input data in order to have it processed. *By reading, you are creating the neural connections for future retrieval.* The more times you read something, the more you reinforce the learning of ideas.

Even if you don't fully understand something on the first pass, *your mind stores much of the material for later recall.*

6. <u>Relax to learn and go into exile</u>. Our bodies respond to an inner clock called biorhythms. Burning the midnight oil works well for some people, but not everyone.

If possible, set aside a particular place to study that is free of distractions. Shut off the television, cell phone, pager and exile your friends and family during your study period.

If you really are bothered by silence, try background music. Light classical music at a low volume has been shown to aid in concentration over other types. Music that evokes pleasant emotions without lyrics are highly suggested. Try just about anything by Mozart. It relaxes you.

7. <u>Use arrows, not highlighters</u>. At best, it's difficult to read a page full of yellow, pink, blue, and green streaks. Try staring at a neon sign for a while and you'll soon see that the horde of colors obscure the message.

A quick note, a brief dash of color, an underline, and an arrow pointing to a particular passage is much clearer than a horde of highlighted words.

8. <u>Budget your study time</u>. Although you shouldn't ignore any of the material, *allocate your available study time in the same ratio that topics may appear on the test.*

Testing Tips:

1. <u>Get smart, play dumb</u>. Don't read anything into the question. Don't make an assumption that the test writer is looking for something else than what is asked. Stick to the question as written and don't read extra things into it.

2. <u>Read the question and all the choices _twice_ before answering the question</u>. You may miss something by not carefully reading, and then re-reading both the question and the answers.

If you really don't have a clue as to the right answer, leave it blank on the first time through. Go on to the other questions, as they may provide a clue as to how to answer the skipped questions.

If later on, you still can't answer the skipped ones . . . *Guess.* The only penalty for guessing is that you *might* get it wrong. Only one thing is certain; if you don't put anything down, you will get it wrong!

3. <u>Turn the question into a statement</u>. Look at the way the questions are worded. The syntax of the question usually provides a clue. Does it seem more familiar as a statement rather than as a question? Does it sound strange?

By turning a question into a statement, you may be able to spot if an answer sounds right, and it may also trigger memories of material you have read.

4. <u>Look for hidden clues</u>. It's actually very difficult to compose multiple-foil (choice) questions without giving away part of the answer in the options presented.

In most multiple-choice questions you can often readily eliminate one or two of the potential answers. This leaves you with only two real possibilities and automatically your odds go to Fifty-Fifty for very little work.

5. <u>Trust your instincts</u>. For every fact that you have read, you subconsciously retain something of that knowledge. On questions that you aren't really certain about, go with your basic instincts. **Your first impression on how to answer a question is usually correct.**

6. <u>Mark your answers directly on the test booklet</u>. Don't bother trying to fill in the optical scan sheet on the first pass through the test.

Just be very careful not to miss-mark your answers when you eventually transcribe them to the scan sheet.

7. <u>Watch the clock</u>! You have a set amount of time to answer the questions. Don't get bogged down trying to answer a single question at the expense of 10 questions you can more readily answer.

I.   **EARTH SCIENCE**

0001 UNDERSTAND CURRENT SCIENTIFIC VIEWS OF THE UNIVERSE

Skill 1.1   **Demonstrating knowledge of the historical progression and characteristics of scientific theories of the formation of the universe**

Two main hypotheses of the origin of the solar system are the **tidal hypothesis** and the **condensation hypothesis**.

The tidal hypothesis proposes that the solar system began with a near collision of the sun and a large star. Some astronomers believe that as these two stars passed each other, the great gravitational pull of the large star extracted hot gases out of the sun. The mass from the hot gases started to orbit the sun, which began to cool then condensing into the nine planets. (Few astronomers support this example).

The condensation hypothesis proposes that the solar system began with rotating clouds of dust and gas. Condensation occurred in the center forming the sun and the smaller parts of the cloud formed the nine planets. (This example is widely accepted by many astronomers).

Two main theories to explain the origins of the universe include: (1) **The Big Bang Theory** and (2) **The Steady-State Theory.**

The Big Bang Theory has been widely accepted by many astronomers. It states that the universe originated from a magnificent explosion spreading mass, matter and energy into space. The galaxies formed from this material as it cooled during the next half-billion years.

The Steady-State Theory is the least accepted theory. It states that the universe is a continuously being renewed. Galaxies move outward and new galaxies replace the older galaxies. Astronomers have not found any evidence to prove this theory.

The future of the universe is hypothesized with the Oscillating Universe Hypothesis. It states that the universe will oscillate or expand and contract. Galaxies will move away from one another and will in time slow down and stop. Then a gradual moving toward each other will again activate the explosion or The Big Bang theory.

Skill 1.2   Analyzing objects in the solar system and the universe based on their characteristics

Astronomers use groups or patterns of stars called **constellations** as reference points to locate other stars in the sky. Familiar constellations include: Ursa Major (also known as the big bear) and Ursa Minor (known as the little bear). Within the Ursa Major, the smaller constellation, The Big Dipper is found. Within the Ursa Minor, the smaller constellation, The Little Dipper is found.

Different constellations appear as the earth continues its revolution around the sun with the seasonal changes.

Magnitude stars are 21 of the brightest stars that can be seen from earth. These are the first stars noticed at night. In the Northern Hemisphere there are 15 commonly observed first magnitude stars.

A vast collection of stars are defined as **galaxies**. Galaxies are classified as irregular, elliptical, and spiral. An irregular galaxy has no real structured appearance; most are in their early stages of life. An elliptical galaxy consists of smooth ellipses, containing little dust and gas, but composed of millions or trillion stars. Spiral galaxies are disk-shaped and have extending arms that rotate around its dense center. Earth's galaxy is found in the Milky Way and it is a spiral galaxy.

Terms related to deep space

A **pulsar** is defined as a variable radio source that emits signals in very short, regular bursts; believed to be a rotating neutron star.

A **quasar** is defined as an object that photographs like a star but has an extremely large redshift and a variable energy output; believed to be the active core of a very distant galaxy.

Black holes are defined as an object that has collapsed to such a degree that light can not escape from its surface; light is trapped by the intense gravitational field.

The **sun** is considered the nearest star to earth that produces solar energy. By the process of nuclear fusion, hydrogen gas is converted to helium gas. Energy flows out of the core to the surface, then radiation escapes into space.

Parts of the sun include: (1) **core:** the inner portion of the sun where fusion takes place, (2) **photosphere:** considered the surface of the sun which produces **sunspots** (cool, dark areas that can be seen on its surface), (3) **chromosphere:** hydrogen gas causes this portion to be red in color (also found here are solar flares (sudden brightness of the chromosphere) and solar prominences (gases that shoot outward from the chromosphere), and (4) **corona:** the transparent area of sun visible only during a total eclipse.

There are eight established planets in our solar system; Mercury, Venus, Earth, Mars, Jupiter, Saturn, Uranus, and Neptune. Pluto was an established ninth planet in our solar system from its discovery in 1930 until the Summer of 2006 when its status was downgraded from planet to one of many dwarf planets and the largest member of a distinct region called the Kuiper belt. The planets are divided into two groups based on distance from the sun. The inner planets include Mercury, Venus, Earth, and Mars. The outer planets include Jupiter, Saturn, Uranus, and Neptune.

Planets

Mercury - the closest planet to the sun. Its surface has craters and rocks. The atmosphere is composed of hydrogen, helium and sodium. Mercury was named after the Roman messenger god.

Venus - has a slow rotation when compared to Earth. Venus and Uranus rotate in opposite directions from the other planets. This opposite rotation is called retrograde rotation. The surface of Venus is not visible due to the extensive cloud cover. The atmosphere is composed mostly of carbon dioxide. Sulfuric acid droplets in the dense cloud cover give Venus a yellow appearance. Venus has a greater greenhouse effect than observed on Earth. The dense clouds combined with carbon dioxide trap heat. Venus was named after the Roman goddess of love.

Earth - considered a water planet with 70% of its surface covered by water. Gravity holds the masses of water in place. The different temperatures observed on earth allow for the different states (solid. Liquid, gas) of water to exist. The atmosphere is composed mainly of oxygen and nitrogen. Earth is the only planet that is known to support life.

Mars - the surface of Mars contains numerous craters, active and extinct volcanoes, ridges, and valleys with extremely deep fractures. Iron oxide found in the dusty soil makes the surface seem rust colored and the skies seem pink in color. The atmosphere is composed of carbon dioxide, nitrogen, argon, oxygen and water vapor. Mars has polar regions with ice caps composed of water. Mars has two satellites. Mars was named after the Roman war god.

Jupiter - largest planet in the solar system. Jupiter has 16 moons. The atmosphere is composed of hydrogen, helium, methane and ammonia. There are white colored bands of clouds indicating rising gas and dark colored bands of clouds indicating descending gases. The gas movement is caused by heat resulting from the energy of Jupiter's core. Jupiter has a Great Red Spot that is thought to be a hurricane type cloud. Jupiter has a strong magnetic field.

Saturn - the second largest planet in the solar system. Saturn has rings of ice, rock, and dust particles circling it. Saturn's atmosphere is composed of hydrogen, helium, methane, and ammonia. Saturn has 20 plus satellites. Saturn was named after the Roman god of agriculture.

Uranus - the second largest planet in the solar system with retrograde revolution. Uranus is a gaseous planet. It has 10 dark rings and 15 satellites. Its atmosphere is composed of hydrogen, helium, and methane. Uranus was named after the Greek god of the heavens.

Neptune - another gaseous planet with an atmosphere consisting of hydrogen, helium, and methane. Neptune has 3 rings and 2 satellites. Neptune was named after the Roman sea god because its atmosphere is the same color as the seas.

Comets, asteroids, and meteors

Astronomers believe that rocky fragments may have been the remains of the birth of the solar system that never formed into a planet. **Asteroids** are found in the region between Mars and Jupiter.

Comets are masses of frozen gases, cosmic dust, and small rocky particles. Astronomers think that most comets originate in a dense comet cloud beyond Pluto. Comet consists of a nucleus, a coma, and a tail. A comet's tail always points away from the sun. The most famous comet, **Halley's Comet,** is named after the person whom first discovered it in 240 B.C. It returns to the skies near earth every 75 to 76 years.

Meteoroids are composed of particles of rock and metal of various sizes. When a meteoroid travels through the earth's atmosphere, friction causes its surface to heat up and it begins to burn. The burning meteoroid falling through the earth's atmosphere is called a **meteor** (also known as a "shooting star").

Meteorites are meteors that strike the earth's surface. A physical example of a meteorite's impact on the earth's surface can be seen in Arizona. The Barringer Crater is a huge meteor crater. There are many other meteor craters throughout the world.

Skill 1.3 **Predicting the effects of gravity (e.g., orbits, tidal forces) on objects in the solar system**

Gravity is a force of attraction between two objects such as the sun and the earth. **Orbits** are the result of balance between the forward motion of a body and the pull of gravity on it from another body. All objects in motion want to stay in forward motion. An object with a lot of mass wants to perpetually move forward, but the gravity of another body pulls it in. This can be compared to a large scale game of tug of war. If either of these forces is changed the orbit will fail, and the body will either crash down onto the object it is orbiting or spin into space.

When an object moves in a circular path, a force must be directed toward the center of the circle in order to keep the motion going. This constraining force is called **centripetal force**. Gravity (of the earth) is the centripetal force that keeps a satellite circling the earth while the sun's gravitational pull keeps the planets in orbit around it. If the sun were not there, the earth would travel in a straight line, but the sun's gravity pulls it out of that straight line. The earth's gravitational pull also acts on the sun to reinforce the orbit. The earth makes one orbit around the sun in 365.25 days. (The earth rotates on its axis once every 24 hours.)

In the same way, the moon orbits the earth due to the pull of the earth's gravitational pull on the moon and the moon's gravitational pull on the earth. The earth's gravity keeps the oceans from falling off the earth and going out into space. **Tides** are changes in the level of the ocean caused by the varying gravitational pull of the moon as it orbits the earth. The moon's gravitational force combines with that of the earth. This interaction produces a common center of gravity between the earth and the moon. This center is called the **barycenter** (the center of gravity between the earth and the moon). As the barycenter rotates around the earth it causes both high tides and low tides. The moon's gravity pulls the water upwards into a bulge on the side of the earth toward the moon and downwards on the far side. This bulge is a daily high tide and the depression on the other side is a daily low tide. The phases of the moon also affect how much gravitation pull the moon exerts with a full moon giving a higher high tide and lower low tide.

The gravitational force of the sun on the oceans is about half that of the moon. When the earth, the moon and the sun are in a line, the forces add together to give the high spring tides twice a month. Seven days later they are at right angles and the forces partially cancel each other, giving a low neap tide.

Skill 1.4   Analyzing the apparent motions of objects in the sky

Because the Earth orbits the sun and rotates on its own axis, bodies in the sky seem to move. While other bodies may be moving as well, the Earth's spin is continuous, and the result is that distant objects appear to move in a predictable manner. For example, we can expect to see certain planets or stars in specific locations at certain times of the year and can chart their 'progress' across the sky.

By learning the constellations, starting with the brightest stars and the clearest formations, we can use the naked eye or a low-power telescope and star maps to follow the patterns and movements. The planets move through (or across) the sky in front of the moving background of stars and can be tracked using maps. Meteor showers, various eclipses and the movements of comets are predictable and are fun to watch.

Skill 1.5   Analyzing the effects of the positions, movements, and interactions of the earth, moon, and sun

The cause and presence of tides are discussed above in Skill 1.3. Other effects of Earth's positioning include the seasons, eclipses, and phases of the moon.

The tilt of the earth's axis allows for the seasonable changes of summer, spring, autumn and winter. As Earth revolves around the sun, the angle of the earth's axis changes relative to the sun. This change results in different amounts of sunlight being received by any one spot throughout the year. We recognize these changes as seasonal variations. The **Summer solstice** occurs when the North Pole is tilted toward the sun on June 21 or 22, providing increased daylight hours for the Northern Hemisphere and shorter daylight hours for the Southern Hemisphere. **Winter solstice** occurs when the South Pole is tilted toward the sun on December 21 or 22, providing shorter daylight hours in the Northern Hemisphere and longer daylight hours in the Southern Hemisphere. The **Spring or Vernal Equinox** occurs on March 20 or 21, when the direct energy from the sun falls on the equator providing equal lengths of day and night hours in both hemispheres. **The Autumn Equinox** occurs on September 22 or 23, again providing the equal amounts of day and night hours in both hemispheres.

Eclipses are defined as the passing of one object into the shadow of another object. A **Lunar eclipse** occurs when the moon travels through the shadow of the earth. A **Solar eclipse** occurs when the moon positions itself between the sun and earth.

The moon keeps the same side of its surface facing the earth at all times. As it orbits the earth, the half facing the sun is illuminated; however, the amount of that side that can be seen from the earth varies, making the moon appear to change shape or phase. From earth, the lighted half of the moon can only be seen when the moon is opposite the sun. The moon is totally dark when it is located in the direction of the sun. The appearances or phases from fully dark to fully light are divided into four groups or quarters, these quarters are designated as (1) new moon, (2) first quarter, (3) full moon, and (4) third quarter. The new moon occurs when the earth, the sun, and the moon are roughly aligned and are invisible to us on the earth. Gradually a bit of the moon becomes visible and each night more of it is illuminated until half is lighted. This is called the first quarter. More and more becomes illuminated until about a week later it appears to be a round moon called a full moon. From "new" to "full" it is said to be waxing. The same side that had become illuminated first begins to darken first so that in about a week only half of the moon is illuminated and is called the third quarter. While the moon is becoming dark, it is said to be waning. Finally the whole moon is dark and is called the new moon.

Skill 1.6   Demonstrating knowledge of the methods (e.g., spectrometry, red-shift analysis) and types of technology used to observe and collect data about the solar system and the universe

Telescope types

Galileo was the first person to use telescopes to observe the solar system. He invented the first refracting telescope. A **refracting telescope** uses lenses to bend light rays to focus the image. The lenses have either concave or convex surfaces, but most are convex. Each telescope may have several lenses. The largest refracting telescope is at the Yerkes Observatory in Wisconsin and has lenses that are 40 inches in diameter.

Sir Isaac Newton invented the **reflecting telescope** using curved mirrors to gather light rays and produce a small focused image. Mirrors are cheaper and can be much larger than the finely ground glass needed for refracting telescopes. The Keck telescopes have 10 meter mirrors and the Gemini telescopes have 8 meter mirrors.

The world's largest telescope is located in Mauna Kea, Hawaii. It uses multiple mirrors to gather light rays.

The **Hubble Space telescope** uses a **single-reflector mirror**. It provides an opportunity for astronomers to observe objects seven times farther away. Even those objects that are 50 times fainter can be viewed better than by any telescope on Earth. There are future plans to make repairs and install new mirrors and other equipment on the Hubble Space telescope.

Refracting and reflecting telescopes are considered **optical telescopes** since they gather visible light and focus it to produce images. A different type of telescope that collects invisible radio waves created by the sun and stars is called a **radio telescope.**

Radio telescopes consist of a reflector or dish with special receivers. The reflector collects radio waves that are created by the sun and stars. Using a radio telescope has many advantages. They can receive signals 24 hours a day, can operate in any kind of weather, and dust particles or clouds do not interfere with their performance. The most impressive aspect of the radio telescope is their ability to detect objects from such great distances in space.

The world's largest radio telescope is located in Arecibo, Puerto Rico. It has a collecting dish antenna of more than 300 meters in diameter.

Spectral analysis

The **spectroscope** is a device or an attachment for telescopes that is used to separate white light into a series of different colors by wavelengths. This series of colors of light is called a **spectrum**. A **spectrograph** can photograph a spectrum. Wavelengths of light have distinctive colors. The color red has the longest wavelength and violet has the shortest wavelength. Wavelengths are arranged to form an **electromagnetic spectrum**. They range from very long radio waves to very short gamma rays. Visible light covers a small portion of the electromagnetic spectrum. Spectroscopes observe the spectra, temperatures, pressures, and also the movement of stars. The movements of stars indicate if they are moving toward, or away, from earth.

If a star is moving toward earth, light waves compress and the wavelengths of light seem shorter. This will cause the entire spectrum to move towards the blue or violet end of the spectrum.

If a star is moving away from earth, light waves expand and the wavelengths of light seem longer. This will cause the entire spectrum to move toward the red end of the spectrum.

Astronomical measurement

The three measurements astronomers use for calculating distances in space are: (1) the **AU or astronomical unit**, (2) **the LY or Light year,** and (3) **the parsec**. It is important to remember that these are measurements of distances not time.

The distance between the earth and the sun is about 150×10^{6} km. This distance is known as an astronomical unit or AU. It is used to measure distances within the solar system; it is not used to measure time.

The distance light travels in one year is a light year (9.5×10^{12} km). This is used to measure distances in space; it does not measure time.
Large distances are measured in parsecs. One parsec equals 3.26 light-years.

There are approximately 63,000 AU's in one light year or,

9.5×10^{12} km/ 150×10^{6} km $= 6.3 \times 10^{4}$ AU

0002 UNDERSTAND THE CHARACTERISTICS AND DISTRIBUTION OF WATER AND ITS ROLE IN EARTH PROCESSES.

Skill 2.1   Analyzing the properties and behavior of water (e.g., polarity, high specific heat capacity, changes in density) in its different states

The unique properties of water are partially responsible for the development of life on Earth. Many of the unique qualities of water stem from the hydrogen bonds that form between the molecules. Hydrogen bonds are particularly strong dipole-dipole interactions that form between the H-atom of one molecule and a Fluorine (F), Oxygen (O), or Nitrogen (N) atom of an adjacent molecule. The partial positive charge on the hydrogen atom is attracted to the partial negative charge on the electron pair of the other atom. The hydrogen bond between two water molecules is shown as the dashed line below:

Note that hydrogen bonding occurs because of the high polarity of the H-O bonds in water. Let's examine how this effect contributes to several of water's unique properties.

1. Density change during freezing
Hydrogen bonds govern the shape of the crystals that form when water freezes. Thus, relatively large spaces of air are present between the water molecules. Since these spaces do not exist when the water is in liquid form, ice is less dense and will float on liquid water.

2. High specific heat/ high heat of vaporization
A molecule the size of water would typically have fairly low specific heat and heat of vaporization. That is, very little energy would be required to trigger a phase change or to heat the water by a given amount. However, extra energy is required to break the network of hydrogen bonds, meaning water has surprisingly high values for both specific heat and heat of vaporization.

3. High surface tension/strong cohesion
Another effect of hydrogen bonding is that water molecules are extremely attracted to one another, as shown in the diagram above. This makes water highly cohesive and produces high surface tension on both droplets of water and large bodies of water. This actually works to prevent massive evaporation of oceans. It also makes raindrops nearly spherical. This same phenomenon is responsible for capillary action.

Skill 2.2 **Analyzing the water cycle (e.g., evaporation, condensation, precipitation, freezing) and its relationship to various atmospheric conditions and hydrologic systems**

The water present now is the water that has been here since our atmosphere formed. Two percent of all the available water is fixed and held in ice or the bodies of organisms. Available water includes surface water (lakes, oceans, and rivers) and ground water (aquifers, wells). 96% of all available water is from ground water. The water located below the surface is called groundwater.

Groundwater provides drinking water for 53% of the population in the United States. Much groundwater is clean enough to drink without any type of treatment. Impurities in the water are filtered out by the rocks and soil through which it flows. However, many groundwater sources are becoming contaminated because of fertilizers, garbage dumping, leaking underground tanks, and various uses of chemicals in manufacturing and farming.

Water that falls to Earth in the form of rain and snow is called **precipitation.** Precipitation is part of a continuous process in which water at the Earth's surface evaporates, condenses into clouds, and returns to Earth. This process is termed the **water cycle**. Relative humidity is the actual amount of water vapor in a certain volume of air compared to the maximum amount of water vapor the same air could hold at a given temperature.

The impacts of altitude upon climatic conditions are primarily related to temperature and precipitation. As altitude increases, climatic conditions become increasingly drier and colder. Solar radiation becomes more severe as altitude increases while the effects of convection forces are minimized. Climatic changes as a function of latitude follow a similar pattern (as a reference, latitude moves either north or south from the equator). The climate becomes colder and drier as the distance from the equator increases. Proximity to land or water masses produce climatic conditions based upon the available moisture. Dry and arid climates prevail where moisture is scarce; lush tropical climates can prevail where moisture is abundant. Climate, as described above, depends upon the specific combination of conditions making up an area's environment. Man impacts all environments by producing pollutants in earth, air, and water. It follows then, that man is a major player in world climatic conditions.

Skill 2.3   Recognizing characteristics and distribution of hydrologic systems on the earth (e.g., rivers, glaciers, groundwater)

Precipitation that soaks into the ground through small pores or openings becomes groundwater. Gravity causes groundwater to move through interconnected porous rock formations from higher to lower elevations. The upper surface of the zone saturated with groundwater is the water table. A swamp is an area where the water table is at the surface. Sometimes the land dips below the water table and these areas fill with water forming lakes, ponds or streams. Groundwater that flows out from underground onto the surface is called a spring.

Permeable rocks filled with water are called **aquifers**. When a layer of permeable rock is trapped between two layers of impermeable rock, an aquifer is formed. Groundwater fills the pore spaces in the permeable rock. Layers of limestone are common aquifers. Groundwater provides drinking water for 53% of the population in the United States and is collected in **reservoirs.**

Precipitation that is collected in running waterbeds becomes a **river.** A river undergoes changes as it flows and erodes the land, passing through different stages of development. There are three main stages in the lifecycle of a river: youth, maturity, and old age.

During the youth stage, a river flows rapidly down mountains and hills. It carries large quantities of coarse-grained sediments. The river cuts and erodes a deep narrow V-shaped channel. The Colorado River is an example of a young river.

A mature river is slow flowing and has less cutting power than a young river. A mature river begins to form curved, S-shaped channels called meanders.

When the slope of a river channel becomes almost flat as it fills with sediment, the water flows very slowly and it is said to be in old age. The lower part of Mississippi River is in this stage.

Glaciers are huge, frozen masses of solid slow-moving ice that form from compacted layers of snow. Gravity moves the glaciers. Glacial ice is the largest reservoir of fresh water on Earth (ocean water is the largest reservoir of total water, but oceans are not fresh water). A temperate glacier is at the melting/freezing point all year. Polar glaciers are always below freezing so the only mass loss is through sublimation (direct ice to atmospheric evaporation). Sub-polar glaciers have a seasonal melting time.

A time period in which glaciers advance over a large portion of a continent is called an ice age. The Ice Age began about 2 -3 million years ago. This age saw the advancement and retreat of glacial ice over millions of years. A moves or flows over land in response to gravity. Glaciers form among high mountains and in other cold regions. There are two main types of glaciers: valley glaciers and continental glaciers. Erosion by valley glaciers is characteristic of U-shaped erosion. They produce sharp peaked mountains such as the Matterhorn in Switzerland. When continental glaciers ride over mountains in their paths, the erosion generally leaves smoothed, rounded mountains and ridges.

Skill 2.4   Interpreting the composition of the world's oceans with respect to location and subsurface topography

Oceans cover over 70% of the Earth's surface. The vast majority of this is salt water and contained in one interconnected body called the World Ocean. However, the World Ocean has traditionally been broken down into several smaller oceans including the Pacific Ocean, the Atlantic Ocean, and Indian Ocean, the Arctic Ocean and the Southern Ocean.

The water itself in these oceans is fairly similar and has a salinity of about 3.5 %. Salinity is the number of grams of dissolved salts in 1,000 grams of sea water. Sodium chloride (NaCl) is the most abundant of the dissolved salts. Smaller quantities of magnesium chloride, magnesium and calcium sulfates, and traces of several other salt elements are found in ocean water. Some areas, such as the Red Sea are considerably saltier and some, such as the Gulf of Finland, are less salty. Salinity is low near river mouths where the ocean mixes with fresh water, and high in areas of high evaporation rates. The density of seawater is about 1020 kg·m^{-3} and the pH varies from 7.5 to 8.4.

The temperature of seawater varies greatly. Certainly latitude plays an important role in this; oceans closer to the poles are nearly always colder. Patterns of wind and sun, however, can create large warming or cooling currents that influence ocean temperatures. Finally, seasonal changes play a role in ocean temperature. The high specific heat of water means that these annual changes are slow, and significant effects are seen mostly in relatively shallow water (such as that near the coast).

The surface of the earth is in constant motion. This motion is the subject of **Plate Tectonics** studies. Major plate separation lines lie along the ocean floors. As these plates separate, molten rock rises, continuously forming new ocean crust and creating new and taller **mountain ridges** under the ocean. The Mid-Atlantic Range, which divides the Atlantic Ocean basin into two nearly equal parts, shows evidence from mapping of these deep-ocean floor changes. **Ocean trenches** are long, elongated narrow troughs or depressions formed where ocean floors collide with another section of ocean floor or continent. The deepest trench in the Pacific Ocean is the Marianas Trench which is about 11 km deep.

Seamounts are formed by underwater volcanoes. Seamounts and **volcanic islands** are found in long chains on the ocean floor. They are formed when the movement of an oceanic plate positions a plate section over a stationary hot spot located deep in the mantle. Magma rising from the hot spot punches through the plate and forms a volcano. The Hawaiian Islands are examples of volcanic island chains. Magma that rises to produce a curving chain of volcanic islands is called an island arc. An example of an island arc is the Lesser Antilles chain in the Caribbean Sea.

The various oceanic habitats are classified largely by temperature, the amount of light received, and distance from shore. Each of the oceanic environments, of course, has its own uniquely adapted organisms. Below is a brief survey of the various geologic structures and the associated oceanic habitat:

Continental shelf: Masses of land extend for a certain distance beneath the surface of the ocean. In certain areas, this continental shelf is extensive; in others it may only extend a few miles. These areas may be known as seas or gulfs and are shallow compared to most ocean regions. They are well lit, well oxygenated, and shallow and so they teem with all types of the most familiar sea creatures.

Abyssal plains: The deep, open ocean is typically underlain with flat or gently sloping abyssal plains. Though some life forms live in the warmer, brighter upper levels of the oceans over abyssal plains, most of the water is deep, cold, dark, poorly oxygenated, and under high pressure. Thus few organisms are present and these areas remain largely unexplored by man.

Estuaries: Estuaries occur where fresh rivers and streams join the ocean. Like the continental shelf, the water here is well lit and shallow. Thus, estuaries are the home for many organisms that are well adapted to the fresh/saline conditions.

Coral reefs: Found in warm, tropical oceans, coral reefs are extensive underwater gardens built on coral skeletons. Their bright, nutrient rich environments are perfect for a wide range of life forms.

Hydrothermal vents: Hydrothermal vents are openings in the Earth's surface, often found beneath deep oceans. They are associated with volcanic activity and release heat and dissolved chemicals. Surprisingly, these areas are often full of life. Certain chemosynthetic bacteria have adapted to these environments and serve as the lowest level on a food chain that may include worms and shellfish.

Skill 2.5   Analyzing the causes and effects of waves, currents, and tides

Waves

The movement of ocean water is caused by the wind, the sun's heat energy, the earth's rotation, the moon's gravitational pull on earth and underwater earthquakes. Most ocean waves are caused by the impact of winds. Wind blowing over the surface of the ocean transfers energy (friction) to the water and causes waves to form. Waves are also formed by seismic activity on the ocean floor. A wave formed by an earthquake is called a seismic sea wave. These powerful waves can be very destructive with wave heights increasing to 30 m or more near the shore.

The crest of a wave is its highest point. The trough of a wave is its lowest point. The distance from wave top to wave top is the wavelength. The wave period is the time between the passing of two successive waves.

Currents

World weather patterns are greatly influenced by ocean surface currents in the upper layer of the ocean. These currents continuously move along the ocean surface in specific directions. **Ocean currents** that flow deep below the surface are called sub-surface currents. These currents are influenced by such factors as the location of landmasses in the current's path and the earth's rotation. **Surface currents** are caused by winds and classified by temperature. Cold currents originate in the Polar regions and flow through surrounding water that is measurably warmer. Those currents with a higher temperature than the surrounding water are called warm currents and can be found near the equator. These currents follow swirling routes around the ocean basins and the equator.

The Gulf Stream and the California Current are the two main surface currents that flow along the coastlines of the United States. The Gulf Stream is a warm current in the Atlantic Ocean that carries warm water from the equator to the northern parts of the Atlantic Ocean. Benjamin Franklin studied and named the Gulf Stream. The California Current is a cold current that originates in the Arctic regions and flows southward along the western coast of the United States. Differences in water density also create ocean currents. Water found near the bottom of oceans is the coldest and the most dense. Water tends to flow from a denser area to a less dense area. These currents that flow because of a difference in the density of the ocean water are called density currents. Water with a higher salinity is more dense than water with a lower salinity. Water that has a salinity different from the surrounding water may form a density current.

Tides

Tides are changes in the level of the ocean caused by the varying gravitational pull of the moon as it orbits the earth. This interaction produces a common center of gravity between the earth and the moon. **Neap tides** are low tides that occur twice a month when the sun, earth, and moon are positioned at right angles to one another. **Spring tides** are abnormally high tides that occur twice a month when the sun, earth, and moon are aligned or positioned in a straight line. The Bay of Fundy in Novia Scotia experiences the largest difference in sea height in the world. Comparing when the tide is highest and lowest, the difference is 45 feet. See Skill 1.3 for more information on Tides.

0003 UNDERSTAND CHARACTERISTICS OF THE ATMOSPHERE AND CLIMATE AND WEATHER

Skill 3.1 **Demonstrating knowledge of the basic composition, structure, and properties of the atmosphere**

Dry air is composed of three basic components; dry gas, water vapor, and solid particles (dust from soil, etc.).

The most abundant dry gases in the atmosphere are:

(N_2)	Nitrogen	78.09 %	makes up about 4/5 of gases in atmosphere
(O_2)	Oxygen	20.95 %	
(AR)	Argon	0.93 %	
(CO_2)	Carbon Dioxide	0.03 %	

The atmosphere is divided into four main layers based on temperature. These layers are labeled Troposphere, Stratosphere, Mesosphere, and Thermosphere.

Troposphere - this layer is the closest to the earth's surface and all weather phenomena occurs here, as it is the layer with the most water vapor and dust. Air temperature decreases with increasing altitude. The average thickness of the Troposphere is 7 miles (11 km).

Stratosphere - this layer contains very little water, clouds within this layer are extremely rare. The Ozone layer is located in the upper portions of the stratosphere. Air temperature is fairly constant but does increase somewhat with height due to the absorption of solar energy and ultra violet rays from the ozone layer.

Mesosphere - air temperature again decreases with height in this layer. It is the coldest layer with temperatures in the range of -100^0 C at the top.

Thermosphere - extends upward into space. Oxygen molecules in this layer absorb energy from the sun, causing temperatures to increase with height. The lower part of the thermosphere is called the Ionosphere. Here charged particles or ions and free electrons can be found. When gases in the Ionosphere are excited by solar radiation, the gases give off light and glow in the sky. These glowing lights are called the Aurora Borealis in the Northern Hemisphere and Aurora Australis in Southern Hemisphere. The upper portion of the Thermosphere is called the Exosphere. Gas molecules are very far apart in this layer. Layers of Exosphere are also known as the Van Allen Belts and are held together by earth's magnetic field.

Skill 3.2   Identifying the processes and patterns of energy transfer in the atmosphere

Energy is transferred in Earth's atmosphere in three ways. Earth gets most of its energy from the sun in the form of waves. This transfer of energy by waves is termed **radiation**. The transfer of thermal energy through matter by actual contact of molecules is called **conduction**. For example, heated rocks and sandy beaches transfer heat to the surrounding air.   The transfer of thermal energy due to air density differences is called **convection**. Convection currents circulate in a constant exchange of cold, dense air for less dense warm air. This helps to create air currents.

Skill 3.3   Comparing heat transfer rates of land and water and analyzing their effects on weather patterns

Most weather on the earth starts with the effects of the sun. The sun's heat warms the earth's atmosphere which causes water to evaporate into the air (forming clouds) and the warm air to rise. As the warm air rises, its temperature begins to drop. Clouds are formed by the cooling moist air when the moisture condenses into droplets. The droplets in the clouds become bigger as more moisture evaporates into the air and then condenses. Eventually, these droplets begin to fall as they are too heavy to stay suspended. Depending upon the temperature when they fall, they can be rain, snow, sleet, or hail.

World weather patterns are greatly influenced by ocean surface currents in the upper layer of the ocean. These currents continuously move along the ocean surface in specific directions. Ocean currents that flow deep below the surface are called sub-surface currents. These currents are influenced by such factors as the location of landmasses in the current's path and the earth's rotation.

Surface currents are caused by winds and are classified by temperature. Cold currents originate in the Polar regions and flow through surrounding water that is measurably warmer. Those currents with a higher temperature than the surrounding water are called warm currents and can be found near the equator. These currents follow swirling routes around the ocean basins and the equator. The Gulf Stream and the California Current are the two main surface currents that flow along the coastlines of the United States. The Gulf Stream is a warm current in the Atlantic Ocean that carries warm water from the equator to the northern parts of the Atlantic Ocean. Benjamin Franklin studied and named the Gulf Stream. The California Current is a cold current that originates in the Arctic regions and flows southward along the west coast of the United States.

Differences in water density also create ocean currents. Water found near the bottom of oceans is the coldest and the densest. Water tends to flow from a denser area to a less dense area. Currents that flow because of a difference in the density of the ocean water are called density currents. Water with a higher salinity is denser than water with a lower salinity. Water that has salinity different from the surrounding water may form a density current.

Skill 3.4   Recognizing characteristics of global and local weather systems and the causes and effects of weather events

The term 'local weather' includes hourly and daily changes in the atmosphere of a region. When we refer to climate we are discussing the average weather for a region over a period of time. Two primary factors discussed in reference to local weather are temperature and precipitation. Climate varies from one place to another because of unequal heating of the Earth's surface. This varied heating of the surface is the result of the unequal distribution of land masses, oceans, and polar ice caps. Differences in area surface temperatures result in pressure gradients. A hot surface heats the air above it and the air expands, which then lowers the air pressure. The result is the creation of wind. This simple system can combine or collide with other simple systems to create complex systems and severe weather. A large scale example is a hurricane, while a small scale example is a coastal breeze.

A **thunderstorm** is a brief, local storm produced by the rapid upward movement of warm, moist air within a cumulo-nimbus cloud. Thunderstorms always produce lightning and thunder, and are accompanied by strong wind gusts and heavy rain or hail.

A severe storm with swirling winds that may reach speeds of hundreds of km per hour is called a **tornado**. Such a storm is also referred to as a "twister". The sky is covered by large cumulo-nimbus clouds and violent thunderstorms; a funnel-shaped swirling cloud may extend downward from a cumulo-nimbus cloud and reach the ground. Tornadoes are storms that leave a narrow path of destruction on the ground.

A swirling, funnel-shaped cloud that **extends** downward and touches a body of water is called a **waterspout.**

Hurricanes are storms that develop when warm, moist air carried by trade winds rotates around a low-pressure "eye". A large, rotating, low-pressure system accompanied by heavy precipitation and strong winds is called a tropical cyclone (better known as a hurricane). In the Pacific region, a hurricane is called a typhoon.

Storms that occur only in the winter are known as blizzards or ice storms. A **blizzard** is a storm with strong winds, blowing snow and frigid temperatures. An **ice storm** consists of falling rain that freezes when it strikes the ground, covering everything with a layer of ice.

Skill 3.5   Analyzing the influence of water on the climate and weather of Georgia

Georgia's climate is heavily influenced by its proximity to the Atlantic Ocean and the Gulf of Mexico. These bodies of water help keep Georgia's climate temperate in general. Additionally, the coastal areas tend to remain cool even in summer, when most of the state becomes intolerably hot. Both these effects are due to the fact the temperature of the ocean and the air above it do not change quickly with the seasons. Thus, sea breezes help to cool the coastal areas and the large body of warm water helps to buffer the low temperature of winter.

The proximity to the oceans also means that Georgia is vulnerable to hurricanes and tropical storms. The risk of these highly damaging storms is highest during the autumn months. Hurricanes and tropical storms can be particularly destructive to those areas directly on the Atlantic coast, but they can also affect areas far inland. The inland areas in Southwest Georgia, for instance, may feel significant effects of the storms that blow in from the Gulf of Mexico.

Of course, like all regions, rainfall is important to the growing season in Georgia, both for agricultural and natural areas. Georgia typically receives 40-50" of rain annually. While other forms of precipitation are rarely seen in southern Georgia, light snow is common a few times a year in the northern mountain regions.

Skill 3.6   Demonstrating knowledge of the methods and types of technology used to observe, measure, and predict climate and weather

Weather instruments that forecast weather include aneroid barometer and the mercury barometer that measure air pressure. The air exerts varying pressures on a metal diaphragm that will read air pressure. The mercury barometer operates when atmospheric pressure pushes on a pool of (mercury) in a glass tube. The higher the pressure the higher up the tube the mercury rises. Anaeroid barometers have a small box inside. The air pressure on this box causes it to change shape, thus moving a needle on a gauge to indicate the air pressure. Normal air pressure reading varies from 28 to 31.

Relative humidity is measured by two kinds of weather instruments, the psychrometer and the hair hygrometer. The psychrometer uses two thermometers, one of which has its bulb covered with a wet cloth. As the cloth dries, the temperature is lowered due to the cooling effect of the evaporation. The temperatures of the two thermometers are then compared on a special chart to find the relative humidity. Relative humidity simply indicates the amount of moisture in the air. Relative humidity is defined as the amount of water vapor in the air compared to the maximum amount of moisture that the air can hold at the same pressure and temperature. Relative humidity is stated as a percentage so the relative humidity can be 0 - 100%.

The sun warms the air, land, and oceans. The land and water retain heat. This heat slowly escapes into the air after the sun disappears. Air temperature (as well as land and water temperatures) can be measured with thermometers. Some thermometers contain red-colored alcohol, older ones contain mercury, and some use a bimetal coil attached to a gauge. The Fahrenheit scale is the older scale and the Celsius scale is now more commonly used.

Meteorologists use an anemometer to measure wind speed and a wind vane to measure wind direction. Wind speed is sometimes rated on a Beaufort scale. Knowing the direction the wind is coming from will help give clues as to how much precipitation and what temperatures to expect.

A rain gauge is an open container with a flat bottom and straight sides which has a scale in inches or centimeters along one side which is clear.

The rain gauge, barometer, anemometer, hygrometer, and thermometer were all invented between 1400 and 1700 and have changed very little in the past 300 years. However, technology has come to play a much larger role in following and predicting weather and weather-related events such as hurricanes and tornadoes. Radar was first used during World War II. It uses radio waves to collect information about precipitation. Doppler radar was invented in the 1950s and adds the ability to measure wind direction and speed. Thus, it is now much simpler to understand and track large storms.

Since the mid-1960s satellites have been taking pictures of the earth. Instruments on board the satellites also capture infrared images of the earth to provide measurements of cloud cover, temperatures at various levels, and water vapor at the various levels. Geostationary satellites (GOES) are located about 22,000 miles above the equator. They circle the earth at the same rate of speed that the earth rotates on its axis, so they stay in the same place relative to the earth all the time. These often provide the images seen on the television weather forecasts. Polar-orbiting satellites (POES) observe the rest of the earth.

Computer-generated and mathematical models have become very complex. They take into account all the physics and geography and past events and trends. Other tools include weather (atmospheric) balloons, weather aircraft, radiosondes, weather rockets, weather ships, ground-based weather stations, and citizen weather watchers who are still an important part of warning systems for tornadoes.

For up-to-date information on all the technology being used to predict weather and track weather events, call a local television station and speak with a meteorologist or call your local Extension Service and speak with an Educator.

Predicting Weather

Local weather is greatly influenced by geography. Meteorologists (scientists who monitor weather conditions) base weather predictions on the chances that certain patterns prevail for each area and certain weather conditions act together in predictable ways.

There are some generalizations that hold true in predicting weather in the United States.

- Most weather conditions move from west to east.
- Along the Atlantic coast, the conditions to the east have great influence.
- Coastlines are more temperate than inland areas. They tend to have cooler summers and warmer winters.
- More moisture falls on the western slopes of mountains than on the eastern slopes.
- Warm air moves up slopes during the day and down slopes at night.
- High altitude areas are usually colder and receive more precipitation than low altitude areas.
- The air above cities is often warmer due to all the concrete and asphalt and people than the surrounding areas. A city can also create an artificial low pressure system.
- In coastal areas, cool air blows inland during the day and out to sea at night.

For ways to predict the weather using easy to make instruments and common, reasonably accurate folklore methods, see:
http://www.wikihow.com/Predict-the-Weather-Without-a-Forecast

Lesson Plans for teachers to analyze data and predict weather can be found at:
http://www.srh.weather.gov/srh/jetstream/synoptic/ll_analyze.htm

For movies of weather events (also PDF activities to use with the movies) see:
http://www.thefutureschannel.com/dockets/realworld/predicting_weather/

Skill 3.7   Knowledge of characteristics of weather, changes in weather, and tools for making weather measurements and predictions

An **air mass** is a large body of air that has similar characteristics of temperature and moisture throughout. Air masses normally form over large flat areas of the earth where air is stagnant for longer periods of time. The air mass can then take on characteristics of the surface below. The *maritime tropical air masses* develop over the subtropical oceans and transport heat and moisture northward in the U.S. The *continental polar air masses* originate over the northern plains of Canada and transport colder, drier air southward.

When an air mass begins to move, it encounters different conditions from its source region. Weather maps are used to show the positions and movement of air masses. When one air mass meets another air mass, a boundary or front is formed. Stormy weather often results.

A **cold front** forms where a cold air mass moves into a warm air mass. The cold air contracts and becomes denser. Therefore, it is heavier than the warm air, so it pushes underneath the warm mass. This creates cumulus clouds and showers or thunderstorms along the cold front. Once the front passes, then the wind changes direction, the skies clear, and the temperature drops.

If the warm mass is moving into a cold air mass, a **warm front** forms. The warmer, less dense air moves up and over the colder air. Early in the process, cirrus clouds may appear. Then there might be stratus clouds and some rain or snow. After the warm front passes, the sky clears, the air pressure rises, and the temperature rises.

If an air mass does not move, it is referred to as a **stationary front**. Precipitation and weak winds can occur.

Air masses moving toward or away from the Earth's surface are called air currents. Air moving parallel to Earth's surface is called **wind**. Weather conditions are generated by winds and air currents carrying large amounts of heat and moisture from one part of the atmosphere to another. Wind speeds are measured by instruments called <u>anemometers</u>.

The wind belts in each hemisphere consist of convection cells that encircle Earth like belts. There are three major wind belts on Earth: (1) trade winds (2) prevailing westerlies, and (3) polar easterlies. Wind belt formation depends on the differences in air pressures that develop in the doldrums, the horse latitudes, and the polar regions. The Doldrums surround the equator. Within this belt heated air usually rises straight up into Earth's atmosphere. The Horse latitudes are regions of high barometric pressure with calm and light winds and the Polar regions contain cold dense air that sinks to the Earth's surface.

As air is heated on a warm day (radiation), the molecules in the air move faster and become further apart. When the molecules are cooled down, they move slower and get closer together. Differences in air pressure help cause winds and affect air masses. Differences in air pressure are shown with lines called isobars on a weather map.

High pressure areas are shown by "H" symbols on weather maps. In these areas, the air pressure is greater than in the surrounding areas. This results in wind, or moving air. In the high pressure area, the air is denser and will move to an area of less density. Therefore, the winds blow away from high pressure areas toward low pressure areas.

Conversely, **low pressure** areas are shown by "L" symbols on weather maps. Since the air is less dense in low pressure areas, the winds blow into the low pressure areas and cause clouds and precipitation.

Air pressure is measured by barometers. Generally, when pressure increases the weather will improve and when pressure decreases the weather will worsen. Winds caused by local temperature changes include sea breezes, and land breezes. **Sea breezes** are caused by the unequal heating of the land and an adjacent, large body of water. Land heats up faster than water. The movement of cool ocean air toward the land is called a sea breeze. Sea breezes usually begin blowing about mid-morning; ending about sunset. A breeze that blows from the land to the ocean or a large lake is called a **land breeze.**

Monsoons are huge wind systems that cover large geographic areas and that reverse direction seasonally. The monsoons of India and Asia are examples of these seasonal winds. They alternate wet and dry seasons. As denser cooler air over the ocean moves inland, a steady seasonal wind called a summer or wet monsoon is produced.

El Niño refers to a sequence of changes in the ocean and atmospheric circulation across the Pacific Ocean. The water around the equator is unusually hot every two to seven years. Trade winds normally blow east to west across the equatorial latitudes, piling warm water into the western Pacific. A huge mass of heavy thunderstorms usually forms in the area and produces vast currents of rising air that displace heat toward the pole. This helps create the strong mid-latitude jet streams. The world's climate patterns are disrupted by this change in location of thunderstorm activity.

Cloud types

Cirrus clouds - White and feathery; high in the sky

Cumulus – thick, white, fluffy

Stratus – layers of clouds cover most of the sky

Nimbus – heavy, dark clouds that represent thunderstorm clouds

Variation on the clouds mentioned above include Cumulo-nimbus and Strato-nimbus.

The air temperature at which water vapor begins to condense is called the **dew point.** All forms of water that falls to the earth is referred to as precipitation. The temperature of the air the water falls through determines the form of precipitation that strikes the earth. If the air temperature is above freezing, the precipitation will most likely be rain and if the air temperature is below freezing, the precipitation will most likely be snow. However, if there are different temperature layers in the air through which the precipitation falls, then it could become sleet or hail. If rain passes through a cold layer, it becomes sleet. Hail is often the result of passing through a series of warmer and colder layers.

Relative humidity is the actual amount of water vapor in a certain volume of air compared to the maximum amount of water vapor this air could hold at a given temperature. Air expands at higher temperatures so it can hold more water vapor at higher temperatures. The amount of humidity in the air affects how fast the water evaporates from the surface of the earth including how fast it evaporates from roads, lakes, and peoples' skin. For example, a temperature of 80 degrees with 60 percent humidity can feel relatively comfortable as perspiration evaporates rather quickly. However, if at 80 degrees the relative humidity is 90 percent, you would feel hot and sticky because perspiration would evaporate very slowly. Humidity is measured with a hygrometer.

0004 UNDERSTAND CHARACTERISTICS OF THE EARTH AND PROCESSES THAT HAVE SHAPED ITS SURFACE

Skill 4.1 Demonstrating knowledge of the earth's structure and composition

The interior of the Earth is divided in to three chemically distinct layers. Starting from the middle and moving towards the surface, these are: the core, the mantle, and the crust. Much of what we know about the inner structure of the Earth has been inferred from various data. Subsequently, there is still some uncertainty about the composition and conditions in the Earth's interior.

Core

The outer core of the Earth begins about 3000 km beneath the surface and is a liquid, though far more viscous than that of the mantle. Even deeper, approximately 5000 km beneath the surface, is the solid inner core. The inner core has a radius of about 1200 km. Temperatures in the core exceed 4000°C. Scientists agree that the core is extremely dense. This conclusion is based on the fact that the Earth is known to have an average density of 5515 kg/m^3 even though the material close to the surface has an average density of only 3000 kg/m^3. Therefore a denser core must exist. Additionally, it is hypothesized that when the Earth was forming, the densest material sank to the middle of the planet. Thus, it is not surprising that the core is about 80% iron. In fact, there is some speculation that the entire inner core is a single iron crystal, while the outer core is a mix of liquid iron and nickel.

Mantle

The Earth's mantle begins about 35 km beneath the surface and stretches all the way to 3000 km beneath the surface, where the outer core begins. Since the mantle stretches so far into the Earth's center, its temperature varies widely; near the boundary with the crust it is approximately 1000°C, while near the outer core it may reach nearly 4000°C. Within the mantle there are silicate rocks, which are rich in iron and magnesium. The silicate rocks exist as solids, but the high heat means they are ductile enough to "flow" over long time scales. In general, the mantle is semi-solid/plastic and the viscosity varies as pressures and temperatures change at varying depths.

Crust

It is not clear how long the Earth has actually had a solid crust; most of the rocks are less than 100 million years, though some are 4.4 billion years old. The crust of the earth is the outermost layer and continues down for between 5 and 70 km beneath the surface. Thin areas generally exist under ocean basins (oceanic crust) and thicker crust underlies the continents (continental crust). Oceanic crust is composed largely of iron magnesium silicate rocks, while continental crust is less dense and consists mainly of sodium potassium aluminum silicate rocks. The crust is the least dense layer of the Earth and so is rich in those materials that "floated" during Earth's formation. Additionally, some heavier elements that bound to lighter materials are present in the crust.

Interactions between the Layers

These layers do not exist as separate entities with little interaction between them. For instance, it is generally believed that a swirling iron-rich liquid in the outer core results in the Earth's magnetic field which is readily apparent on the surface. Heat also moves out from the core to the mantle and crust. The core still retains heat from the formation of the Earth and additional heat is generated by the decay of radioactive isotopes. While most of the heat in our atmosphere comes from the sun, radiant heat from the core does warm oceans and other large bodies of water.

There is also a great deal of interaction between the mantle and the crust. The slow convection of rocks in the mantle is responsible for the shifting of tectonic plates on the crust. Matter can also move between the layers as occurs during the rock cycle. Within the rock cycle, igneous rocks are formed when magma escapes from the mantle as lava during volcanic eruption. Rocks may also be forced back into the mantle, where the high heat and pressure recreate them as metamorphic rocks.

Skill 4.2   Analyzing processes of the rock cycle and classifying rocks based on their characteristics

The three major subdivisions of rocks are sedimentary, metamorphic and igneous. By volume, more than two-thirds of the earth's crust is composed of igneous rock and about one-fourth is metamorphic. While sedimentary rock only makes up about 8% of the earth's crust, it is the most likely to be exposed at the surface.

The Rock Cycle

The rock cycle is a continuous process of creating and destroying the surface of the earth. It is not obvious to the average person, nor is it visible to the naked eye. Matter from the earth's crust is changed from one form to another but never created or lost. The rock cycle represents the alteration of rock-forming minerals above and below the earth's surface.

Molten magma is uplifted from inside the earth. It oozes or spews onto the surface. If it is extruded on the surface, it is lava. It may cool above the surface and give rocks with fine textures or it may cool more slowly below the surface and give rocks with larger crystals. These igneous rocks can remain as igneous rocks to be exposed at some later date to weathering and erosion or they can be re-melted when exposed to intense heat to become magma again or they can be changed into metamorphic rock.

When igneous rock is exposed to weather and erosion, it forms sediment as a result of freezing and thawing or rain water soaking through pores. Acid seeps into the rock. Sun-warmed air expands the rock and cold air contracts it. Eventually, pieces begin to chip off and break into smaller and smaller pieces until it is a pile of rubble and dust called sediment.

The sediment can be carried away by rain and wind. It becomes deposited elsewhere. Lithification takes place to create sedimentary rocks. In the process, fossils may be formed or deposited in the sediment to become part of the sedimentary rock.

Sedimentary rocks can then be broken down again into sediment, or exposed to intense heat and/or pressure to become metamorphic rock, or melted into magma by even more intense heat. Plate tectonic movements can push sedimentary rock into the mantle to be re-melted into magma and recycled.

Metamorphic rocks have been altered by exposure to great heat and/or pressure. This can come from many heavy layers on top of the metamorphic rocks or the collision of lithospheric plates. If the material melts completely, the rock becomes magma again.

Lithification of sedimentary rocks

Sedimentary rocks are formed from fragments of preexisting rocks that are transported and deposited by wind, water, or glaciers. They can also be formed by the precipitation of solids or the evaporation of water from a solution. When fluid sediments are transformed into solid sedimentary rocks, the process is known as **lithification**. One very common process affecting sediments is compaction where the weights of overlying materials compress and compact the deeper sediments. The compaction process leads to cementation. **Cementation** is when sediments are converted to sedimentary rock.

Factors in crystallization of igneous rocks

Igneous rocks can be classified according to their texture, their composition, and the way they were formed.

Molten rock is called magma. When molten rock pours out onto the surface of Earth, it is called lava.

As magma cools, the elements and compounds begin to form crystals. The slower the magma cools, the larger the crystals grow. Rocks with large crystals are said to have a coarse-grained texture. Granite is an example of a coarse grained rock. Rocks that cool rapidly before any crystals can form have a glassy texture such as obsidian, also commonly known as volcanic glass.

There are two general types of extrusive (volcanic) magma. Basaltic magma comes directly from the mantle. It flows easily and forms thin layers that cover large areas like the ocean floor. Granitic magma is thick and stiff. Volcanoes of this material are often explosive and blow out large amounts of ash and huge pieces of rock.

Intrusive (plutonic) igneous rock was formed below the earth's surface as molten material pushed its way upward through the rocks, cutting across or squeezing between them and solidifying before reaching the surface. This type of rock typically has large crystals. It appears in irregular shapes.

Metamorphic rocks are formed by high temperatures and great pressures. The process by which the rocks undergo these changes is called metamorphism. The outcome of metamorphic changes include deformation by extreme heat and pressure, compaction, destruction of the original characteristics of the parent rock, bending and folding while in a plastic stage, and the emergence of completely new and different minerals due to chemical reactions with heated water and dissolved minerals.

Metamorphic rocks are classified into two groups, foliated (leaflike) rocks and unfoliated rocks. Foliated rocks consist of compressed, parallel bands of minerals, which give the rocks a striped appearance. Examples of such rocks include slate, schist, and gneiss. Unfoliated rocks are not banded and examples of such include quartzite, marble, and anthracite rocks.

Minerals are natural, non-living solids with a definite chemical composition and a crystalline structure. **Ores** are minerals or rock deposits that can be mined easily and a needed or useful metal or nonmetal must be able to be extracted for a profit. The feldspars of the continental crust contain aluminum but it cannot be easily extracted so they are considered rock-forming rather than ore minerals. On the other hand, some ores are pure metals such gold (Au) nuggets. Many ore minerals are oxides (O) combined with a metal such as magnetite (Fe_3O_4) or hematite (Fe_2O_3). Other ore minerals are sulfides (combined with sulfur) such as iron pyrites (FeS_2) or galena (PbS).

Gems are rare minerals which are considered beautiful as well as durable. The rarest and most valuable gems are called precious stones. These include diamonds, emeralds, alexandrites, and aquamarines. Semiprecious stones are not as rare, but are beautiful and durable. They include amethysts, zircon, garnets, turquoise, jade, malachite and opals. Most birthstones and stones used in jewelry are precious or semiprecious gem stones.

Rocks are earth materials made of one or more minerals. A **Rock Facies** is a rock group that differs from comparable rocks (as in composition, age or fossil content).

Characteristics by which minerals are classified

There are over 3000 minerals in Earth's crust. Only about 20 minerals are common and 10 make up 90% of the earth's crust. Minerals are classified by composition. Minerals can be a single element like iron (Fe) or gold (Au) or a combination of elements. The major groups of minerals are metals (positive ions) combined with polyatomic ions such as the silicates (SiO_4^{4-}), carbonates (CO_3^{2-}), oxides (O^{2-}), sulfides (S^{2-}), sulfates (SO_4^{2-}), and halides (H^{1-}). The largest group of minerals is the silicates. Silicates are made of silicon, oxygen, and one or more other elements. Minerals must adhere to five criteria. A mineral must (1) be non-living, (2) be formed in nature, (3) be solid in form, (4) have atoms which form a crystalline pattern, (5) have a chemical composition that is fixed within narrow limits.

The key to understanding the structure of a mineral is to study the arrangement of its ions. That arrangement depends upon the electrical charges between the ions and their relative sizes. Minerals are distinguished from each other based on their chemical composition and crystalline structure, luster, hardness, weight (specific gravity), color, fluorescence, magnetism, solubility, cleavage, and radioactivity.

Two or more minerals can have the same **chemical composition** but different internal structures. For example, graphite and diamond are chemically the same, but their crystalline structures are dramatically different. Crystals do not normally grow evenly; one face grows faster than another so the final crystal looks nothing like the theoretical type for that mineral.

The **crystalline structure** may or may not be evident from the surface of the mineral, but it will be more evident based on its breakage (fracture or cleavage) which will expose its sides or crystal faces. Planes of weakness in the crystal lattice reveal themselves in the tendency for the crystal to split in a certain direction. For example, mica has silicate molecules arranged in flat sheets so it flakes away like the leaves of a book while calcite breaks apart into perfect mini-crystals. Both are examples of cleavage. However, some minerals fracture by breaking along an irregular surface. Fracture surfaces can be conchoidal or curved like the inside of a shell, hackly, splintery, or fibrous.

The number of possible internal arrangements is limited. There are six basic crystalline structures.

1. Cubic or isometric – 3 axes at right angles to one another and all of the same length. Examples: Iron pyrites (FeS_2), garnet
2. Tetragonal – 3 axes, all at right angles to one another, two of which have the same length. Examples: Copper iron sulfide ($CuFeS_2$), zircon
3. Hexagonal – 4 axes, three of which have the same length, at 120° to one another and at right angles to the fourth. Examples: beryllium silicate ($Be_3Al_2(SiO_3)_6$), quartz, calcite, dolomite, hematite
4. Orthorhombic – 3 axes all at right angles to one another but of unequal lengths. Examples: Topaz ($Al_2F_2SiO_4$), olivine
5. Monoclinic – 3 axes, one not at right angles to the others, of unequal lengths. Examples: Pyroxene mica clay, arthoclase, gypsum
6. Triclinic – 3 axes, none of which is at right angles to any of the others, and all of which are different lengths. Examples: albite feldspar

Hardness is another method for identifying minerals. The Mohs Scale is used by geologists to assess hardness. It is a scale of "1" to "10" with "1" being the softest (talc) and "10" being the hardest (diamond). A simplified look at the Mohs Scale of hardness shows:

1	Talc	Fingernail can scratch
2	Gypsum	
3	Calcite	Penny can scratch
4	Fluorite	Knife or glass can scratch
5	Apatite	
6	Orthoclase	
7	Quartz	File can scratch
8	Topaz	Quartz can scratch
9	Ruby, Sapphire	
10	Diamond	

The **colors** of minerals aid in their identification. Streak refers to the color of the mineral when it has been made into powder. A *streak* can be produced by scratching a piece of the mineral across the surface of an unglazed white ceramic tile (unless the mineral is harder than the tile). For example, pyrite which is known as "fool's gold" has a yellowish, metallic colored streak while feldspars have a white or pinkish streak.

Luster describes how the freshly broken (fractured) surface of a mineral reflects light. Many ore minerals have a *metallic* luster. Most of the silicate minerals have a glassy luster called *vitreous*. *Resinous* luster is similar to plastic in appearance. *Silky* luster (example: gypsum) is typical of minerals that are formed of thick masses of very fine hair-like crystals. *Pearly* luster looks like pearls.

Specific gravity is the weight of a mineral compared to the weight of an equal volume of water. The mass of the sample is taken. It is then immersed in a known (measured) amount of water and the measure of the water plus the sample is taken. By subtracting the original known amount of water from that, the amount of displaced water can be found. Dividing the weight of the mineral by the weight of the displaced water gives the specific gravity. A mineral with a specific gravity of 3 or more is noticeably heavy, so galena with a specific gravity of 7.6 would be "very heavy."

Skill 4.3 Describing the characteristics of soil and its formation

Soils are composed of particles of sand, clay, various minerals, tiny living organisms, and humus, plus the decayed remains of plants and animals. Soils are divided into three classes according to their texture. These classes are sandy soils, clay soils, and loamy soils.

Sandy soils are gritty, and their particles do not bind together firmly. Sandy soils are porous- water passes through them rapidly. Sandy soils do not hold much water.

Clay soils are smooth and greasy, their particles bind together firmly. Clay soils are moist and usually do not allow water to pass through easily.

Loamy soils feel somewhat like velvet and their particles clump together. Loamy soils are made up of sand, clay, and silt. Loamy soils holds water but some water can pass through.

In addition to three main classes, soils are further grouped into three major types based upon their composition. These groups are pedalfers, pedocals, and laterites.

Pedalfers form in the humid, temperate climate of the eastern United States. Pedalfer soils contain large amounts of iron oxide and aluminum-rich clays, making the soil a brown to reddish brown color. This soil supports forest type vegetation.

Pedocals are found in the western United States where the climate is dry and temperate. These soils are rich in calcium carbonate. This type of soil supports grasslands and brush vegetation.

Laterites are found where the climate is wet and tropical. Large amounts of water flows through this soil. Laterites are red-orange soils rich in iron and aluminum oxides. There is little humus and this soil is not very fertile.

Skill 4.4 Applying knowledge of the theory of plate tectonics and analyzing evidence that supports the theory

Data obtained from many sources led scientists to develop the theory of plate tectonics. This theory is the most current model that explains not only the movement of the continents, but also the changes in the earth's crust caused by internal forces.

Plates are rigid blocks of earth's crust and upper mantle. These rigid solid blocks make up the lithosphere. The earth's lithosphere is broken into nine large sections and several small ones. These moving slabs are called plates. The major plates are named after the continents they are "transporting."

The plates float on and move with a layer of hot, plastic-like rock in the upper mantle. Geologists believe that the heat currents circulating within the mantle cause this plastic zone of rock to slowly flow, carrying along the overlying crustal plates.

Movement of these crustal plates creates areas where the plates diverge as well as areas where the plates converge. A major area of divergence is located in the Mid-Atlantic. Currents of hot mantle rock rise and separate at this point of divergence creating new oceanic crust at the rate of 2 to 10 centimeters per year. Convergence is when the oceanic crust collides with either another oceanic plate or a continental plate. The oceanic crust sinks forming an enormous trench and generating volcanic activity. Convergence also includes continent to continent plate collisions. When two plates slide past one another a transform fault is created.

These movements produce many major features of the earth's surface, such as mountain ranges, volcanoes, and earthquake zones. Most of these features are located at plate boundaries, where the plates interact by spreading apart, pressing together, or sliding past each other. These movements are very slow, averaging only a few centimeters a year.

Boundaries form between spreading plates where the crust is forced apart in a process called **rifting**. Rifting generally occurs at mid-ocean ridges. Rifting can also take place within a continent, splitting the continent into smaller landmasses that drift away from each other, thereby forming an ocean basin between them. The Red Sea is a product of rifting. As the seafloor spreading takes place, new material is added to the inner edges of the separating plates. In this way the plates grow larger, and the ocean basin widens. This is the process that broke up the super continent Pangaea and created the Atlantic Ocean.

Boundaries between plates that are colliding are zones of intense crustal activity. When a plate of ocean crust collides with a plate of continental crust, the more dense oceanic plate slides under the lighter continental plate and plunges into the mantle. This process is called **subduction**, and the site where it takes place is called a subduction zone. A subduction zone is usually seen on the sea-floor as a deep depression called a trench.

The crustal movement which is identified by plates sliding sideways past each other produces a plate boundary characterized by major faults that are capable of unleashing powerful earth-quakes. The San Andreas Fault forms such a boundary between the Pacific Plate and the North American Plate.

Skill 4.5   Identifying major geological features and analyzing the processes that form and change these features (e.g., tectonic movements, erosion, deposition)

Orogeny is the term given to mountain building that occurs naturally.

A mountain is terrain that has been raised high above the surrounding landscape by volcanic action, or some form of tectonic plate collisions. The plate collisions could be intercontinental or ocean floor collisions with a continental crust (subduction). The physical composition of mountains would include igneous, metamorphic, or sedimentary rocks; some may have rock layers that are tilted or distorted by plate collision forces.

There are many different types of mountains. The physical attributes of a mountain range depends upon the angle at which plate movement thrust layers of rock to the surface. Many mountains (Adirondacks, Southern Rockies) were formed along high angle faults.

Folded mountains (Alps, Himalayas) are produced by the folding of rock layers during their formation. The Himalayas are the highest mountains in the world and contain Mount Everest, which rises almost 9 km above sea level. The Himalayas were formed when India collided with Asia. The movement that created this collision is still in process at the rate of a few centimeters per year.

Fault-block mountains (Utah, Arizona, and New Mexico) are created when plate movement produces tension forces instead of compression forces. The area under tension produces normal faults and rock along these faults is displaced upward.

Dome mountains are formed as magma tries to push up through the crust but fails to break the surface. Dome mountains resemble a huge blister on the earth's surface.

Upwarped mountains (Black Hills of S.D.) are created in association with a broad arching of the crust. They can also be formed by rock thrust upward along high angle faults.

Volcanism is the term given to the movement of magma through the crust and its emergence as lava onto the earth's surface. Volcanic mountains are built up by successive deposits of volcanic materials.

An active volcano is one that is presently erupting or building to an eruption. A dormant volcano is one that is between eruptions but still shows signs of internal activity that might lead to an eruption in the future. An extinct volcano is said to be no longer capable of erupting. Most of the world's active volcanoes are found along the rim of the Pacific Ocean, which is also a major earthquake zone. This curving belt of active faults and volcanoes is often called the Ring of Fire.

The world's best known volcanic mountains include: Mount Etna in Italy and Mount Kilimanjaro in Africa. The Hawaiian Islands are actually the tops of a chain of volcanic mountains that rise from the ocean floor.

There are three types of volcanic mountains: shield volcanoes, cinder cones and composite volcanoes.

Shield Volcanoes are associated with quiet eruptions. Lava emerges from the vent or opening in the crater and flows freely out over the earth's surface until it cools and hardens into a layer of igneous rock. A repeated lava flow builds this type of volcano into the largest volcanic mountain. Mauna Loa found in Hawaii, is the largest volcano on earth.

Cinder Cone Volcanoes are associated with explosive eruptions as lava is hurled high into the air in a spray of droplets of various sizes. These droplets cool and harden into cinders and particles of ash before falling to the ground.

The ash and cinder pile up around the vent to form a steep, cone-shaped hill called the cinder cone. Cinder cone volcanoes are relatively small but may form quite rapidly.

Composite Volcanoes are described as being built by both lava flows and layers of ash and cinders. Mount Fuji in Japan, Mount St. Helens in Washington, USA and Mount Vesuvius in Italy are all famous composite volcanoes.

Mechanisms of producing mountains

Mountains are produced by different types of mountain-building processes. Most major mountain ranges are formed by the processes of folding and faulting.

Folded Mountains are produced by the folding of rock layers. Crustal movements may press horizontal layers of sedimentary rock together from the sides, squeezing them into wavelike folds. Up-folded sections of rock are called anticlines; down-folded sections of rock are called synclines. The Appalachian Mountains are an example of folded mountains with long ridges and valleys in a series of anticlines and synclines formed by folded rock layers.

Faults are fractures in the earth's crust which have been created by either tension or compression forces transmitted through the crust. These forces are produced by the movement of separate blocks of crust.

Faultings are categorized on the basis of the relative movement between the blocks on both sides of the fault plane. The movement can be horizontal, vertical or oblique.

A dip-slip fault occurs when the movement of the plates is vertical and opposite. The displacement is in the direction of the inclination, or dip, of the fault. Dip-slip faults are classified as normal faults when the rock above the fault plane moves down relative to the rock below.

Reverse faults are created when the rock above the fault plane moves up relative to the rock below. Reverse faults having a very low angle to the horizontal are also referred to as thrust faults.

Faults in which the dominant displacement is horizontal movement along the trend or strike (length) of the fault are called **strike-slip faults**. When a large strike-slip fault is associated with plate boundaries it is called a **transform fault**. The San Andreas Fault in California is a well-known transform fault.

Faults that have both vertical and horizontal movement are called **oblique-slip faults**.

When lava cools, igneous rock is formed. This formation can occur either above ground or below ground.

Intrusive rock includes any igneous rock that was formed below the earth's surface. Batholiths are the largest structures of intrusive type rock and are composed of near-granite materials; they are the core of the Sierra Nevada Mountains.

Extrusive rock includes any igneous rock that was formed at the earth's surface.

Dikes are old lava tubes formed when magma entered a vertical fracture and hardened. Sometimes magma squeezes between two rock layers and hardens into a thin horizontal sheet called a **sill**. A **laccolith** is formed in much the same way as a sill, but the magma that creates a laccolith is very thick and does not flow easily. It pools and forces the overlying strata creating an obvious surface dome.

A **caldera** is normally formed by the collapse of the top of a volcano. This collapse can be caused by a massive explosion that destroys the cone and empties most if not all of the magma chamber below the volcano. The cone collapses into the empty magma chamber forming a caldera.

An inactive volcano may have magma solidified in its pipe. This structure, called a volcanic neck, is resistant to erosion and today may be the only visible evidence of the past presence of an active volcano.

When lava cools, igneous rock is formed. This formation can occur either above ground or below ground.

Glaciation

A continental glacier covered a large part of North America during the most recent ice age. Evidence of this glacial coverage remains as abrasive grooves, large boulders from northern environments dropped in southerly locations, glacial troughs created by the rounding out of steep valleys by glacial scouring, and the remains of glacial sources called **cirques** that were created by frost wedging the rock at the bottom of the glacier. Remains of plants and animals found in warm climate have been discovered in the moraines and out wash plains help to support the theory of periods of warmth during the past ice ages.

The Ice Age began about 2 -3 million years ago. This age saw the advancement and retreat of glacial ice over millions of years. Theories relating to the origin of glacial activity include Plate Tectonics, where it can be demonstrated that some continental masses, now in temperate climates, were at one time blanketed by ice and snow. Another theory involves changes in the earth's orbit around the sun, changes in the angle of the earth's axis, and the wobbling of the earth's axis. Support for the validity of this theory has come from deep ocean research that indicates a correlation between climatic sensitive micro-organisms and the changes in the earth's orbital status.

About 12,000 years ago, a vast sheet of ice covered a large part of the northern United States. This huge, frozen mass had moved southward from the northern regions of Canada as several large bodies of slow-moving ice, or glaciers. A time period in which glaciers advance over a large portion of a continent is called an ice age. A glacier is a large mass of ice that moves or flows over the land in response to gravity. Glaciers form among high mountains and in other cold regions.

There are two main types of glaciers: valley glaciers and continental glaciers. Erosion by valley glaciers is characteristic of U-shaped erosion. They produce sharp peaked mountains such as the Matterhorn in Switzerland. Erosion by continental glaciers often rides over mountains in their paths leaving smoothed, rounded mountains and ridges.

Erosion is the inclusion and transportation of surface materials by another moveable material, usually water, wind, or ice. The most important cause of erosion is running water. Streams, rivers, and tides are constantly at work removing weathered fragments of bedrock and carrying them away from their original location.

A stream erodes bedrock by the grinding action of the sand, pebbles and other rock fragments. This grinding against each other is called abrasion.

Streams also erode rocks by dissolving or absorbing their minerals. Limestone and marble are readily dissolved by streams.

The breaking down of rocks at or near to the earth's surface is known as **weathering**. Weathering breaks down these rocks into smaller and smaller pieces. There are two types of weathering: physical weathering and chemical weathering.

Physical weathering is the process by which rocks are broken down into smaller fragments without undergoing any change in chemical composition. Physical weathering is mainly caused by the freezing of water, the expansion of rock, and the activities of plants and animals.

Frost wedging is the cycle of daytime thawing and refreezing at night. This cycle causes large rock masses, especially the rocks exposed on mountain tops, to be broken into smaller pieces.

The peeling away of the outer layers from a rock is called exfoliation. Rounded mountain tops are called exfoliation domes and have been formed in this way.

Chemical weathering is the breaking down of rocks through changes in their chemical composition. An example would be the change of feldspar in granite to clay. Water, oxygen, and carbon dioxide are the main agents of chemical weathering. When water and carbon dioxide combine chemically, they produce a weak acid that breaks down rocks.

Skill 4.6   Recognizing characteristics of fossils and their formation

A fossil is the remains or trace of an ancient organism that has been preserved naturally in the Earth's crust. Sedimentary rocks usually are rich sources of fossil remains. Those fossils found in layers of sediment were embedded in the slowly forming sedimentary rock strata. The oldest fossils known are the traces of 3.5 billion year old bacteria found in sedimentary rocks. Few fossils are found in metamorphic rock and virtually none found in igneous rocks. The magma is so hot that any organism trapped in the magma is destroyed.

The fossil remains of a woolly mammoth embedded in ice were found by a group of Russian explorers. However, the best-preserved animal remains have been discovered in natural tar pits. When an animal accidentally fell into the tar, it became trapped sinking to the bottom. Preserved bones of the saber-toothed cat have been found in tar pits.

Prehistoric insects have been found trapped in ancient amber or fossil resin that was excreted by some extinct species of pine trees.

Fossil molds are the hollow spaces in a rock previously occupied by bones or shells. A fossil cast is a fossil mold that fills with sediments or minerals that later hardens forming a cast.

Fossil tracks are the imprints in hardened mud left behind by birds or animals.

Skill 4.7   Interpreting evidence (e.g., fossils, trapped gases) of the changing surface and climate of the earth

Earth's history extends over more than four billion years and is reckoned in terms of a scale. Paleontologists who study the history of the Earth have divided this huge period of time into four large time units called Eons. Eons are divided into smaller units of time called eras. An Era refers to a time interval in which particular plants and animals were dominant, or present in great abundance. The end of an Era is most often characterized by (1) a general uplifting of the crust, (2) the extinction of the dominant plants or animals, and (3) the appearance of new life forms. Eras are further divided into Periods which are further refined into Epochs.

Methods of geologic dating

Estimates of the Earth's age have been made possible with the discovery of **radioactivity** and the invention of instruments that can measure the amount of radioactivity in rocks. The use of radioactivity to make accurate determinations of Earth's age is called Absolute Dating. This process depends upon comparing the amount of radioactive material in a rock with the amount that has decayed into another element. Studying the radiation given off by atoms of radioactive elements is the most accurate method of measuring the Earth's age. These atoms are unstable and are continuously breaking down or undergoing decay. The radioactive element that decays is called the parent element. The new element that results from the radioactive decay of the parent element is called the daughter element.

The time required for one half of a given amount of a radioactive element to decay is called the half-life of that element or compound. Geologists commonly use Carbon Dating to calculate the age of a fossil substance.

Inferring the history of an area using geologic evidence

The determination of the age of rocks by cataloging their composition has been outmoded since the middle 1800s. Today a sequential history can be determined by the fossil content (principle of fossil succession) of a rock system as well as its superposition within a range of systems. This classification process was termed stratigraphy and permitted the construction of a Geologic Column in which rock systems are arranged in their correct chronological order. The Law of Superposition states that in sedimentary rock each layer is older than the one above it and younger than the one below.

Specific fossils, each of which lived at a specific time and for a limited period of time, are used as representative of those time periods. These fossils are called index fossils. Because fossils actually record the slow but progressive development of life, scientists can use them to identify rocks of the same age throughout the world.

Annual variations such as growth rings in tree trunks also aid in telling the ages of areas. In and near most glacial ice fields are seasons of melting and freezing which are sharply marked. Lakes and ponds in these areas hold preserved deposits.

Principles of catastrophism and uniformitarianism

Uniformitarianism - is a fundamental concept in modern geology. It simply states that the physical, chemical, and biological laws that operated in the geologic past operate in the same way today. The forces and processes that we observe presently shaping our planet have been at work for a very long time. This idea is commonly stated as "the present is the key to the past."

Catastrophism - the concept that the earth was shaped by catastrophic events of a short term nature.

0005 UNDERSTAND THE EARTH'S NATURAL RESOURCES

Skill 5.1 **Identifying types and characteristics of renewable and nonrenewable resources and analyzing factors that influence how they are used**

A **renewable resource** is one that is replaced naturally. Living renewable resources would be plants and animals. Plants are renewable because they grow and reproduce. Sometimes renewal of the resource doesn't keep up with the demand. Such is the case with trees. Since the housing industry uses lumber for frames and homebuilding they are often cut down faster than new trees can grow. Now there are specific tree farms. Special methods allow trees to grow faster.

A second renewable resource is animals. They renew by the process of reproduction. Some wild animals need protection on refuges. As the population of humans increases resources are used faster. Cattle are used for their hides and for food. Some animals like deer are killed for sport. Each state has an environmental protection agency with divisions of forest management and wildlife management.

Non-living renewable resources would be water, air, and soil. Water is renewed in a natural cycle called the water cycle. Air is a mixture of gases. Oxygen is given off by plants and taken in by animals that in turn expel the carbon dioxide that the plants need. Soil is another renewable resource. Fertile soil is rich in minerals. When plants grow they remove the minerals and make the soil less fertile. Chemical treatments are one way or renewing the composition. It is also accomplished naturally when the plants decay back into the soil. The plant material is used to make compost to mix with the soil.

Nonrenewable resources are not easily replaced in a timely fashion. Minerals are nonrenewable resources. Quartz, mica, salt and sulfur are some examples. Mining depletes these resources so society may benefit. Glass is made from quartz, electronic equipment from mica, and salt has many uses. Sulfur is used in medicine, fertilizers, paper, and matches.

Metals are among the most widely used nonrenewable resource. Metals must be separated from the ore. Iron is our most important ore. Gold, silver and copper are often found in a more pure form called native metals.

Skill 5.2   Recognizing the sun as the major source of energy on the earth's surface and analyzing its relationship to wind and water energy

The vast majority of energy on the Earth's surface is derived from sunlight. However, sunlight is attenuated by the Earth's atmosphere, so that not all this solar energy reaches the planet's surface. Specifically, about 1300 watts are delivered per square meter of Earth, but only about 1000 watts actually reach the surface.

It is easy for us to understand how sunlight warms the land and water on the surface of the Earth. We are similarly familiar with the capture of sunlight by solar cells, which then can be used for heating or electricity. However, it is important to understand that sunlight is the basis for many of our other sources of power. This is because sunlight is used to drive photosynthesis, which is the major method by which carbon is fixed by living things. The energy harnessed by plants is used to fuel all heterotrophs further up the food chain. When these life forms (plant or animal) die, they may ultimately be converted to fossil fuels. Thus the petroleum, oil, and other fossil fuels we use as a major power source all originally derived their energy from sunlight.

The sun and the energy it transfers to the Earth also influence the movement of air and water (i.e., the winds and ocean currents). This is a result of the fact that different areas of the Earth's surface receive varying amounts of energy from the sunlight. The movements of warmer and cooler masses of air or water are the source of wind and water currents. Note that this means that wind and hydrologic energy also have their origin in energy from the sun.

Finally, it is important to recognize that the sun is a star with its own strong and ever-changing magnetic field. These changes in the field are responsible for solar activity such as sunspots, solar flares, and solar wind. These changes in turn affect Earth's climate and sometimes increase or decrease the solar energy that reaches the entire Earth or certain areas. Fluctuations in the solar cycles can also lead to especially low or high temperatures. For instance, during the 17th and 18th centuries there was a period in which very few sunspots were seen. This lasted for about 70 years and coincided with the "Little Ice Age" in Europe, an era of unusually cold temperature throughout the continent. It is not only the extended period of changes in solar activity that effect climate and weather on Earth. For example, it has been observed that stratospheric winds near the equator blow in different directions, depending on changes in the solar cycle.

Skill 5.3   Recognizing the sources of geothermal energy

Geothermal energy originates deep within the Earth. Both the core and mantle of the Earth are extremely hot and this heat flows up to the surface. This flow of heat is about 45×10^{12} Watts. The core of the Earth is located 3,200 to 4,000 miles beneath the surface of the Earth and is believed to have a temperature of 5,000 to 6,000°C. The core is extremely dense, under high pressure, and thought to be largely solid. The mantle is also quite hot, but both pressure and temperature decrease as one moves away from the core toward the crust. The mantle is solid/plastic with slow moving, semi-solid large pieces of rock. The generation of heat within the Earth's interior is not completely understood, though many mechanisms have been postulated by making inferences from what is known about the Earth's core. It is likely that multiple phenomena are responsible and include the following:

1. Nuclear decay of naturally radioactive elements within the core; this is believed to be largest source of geothermal energy.

2. Residual heat from the formation of the planet; this theory is supported by the fact that the Earth's interior appears to be slowly cooling.

3. The Earth's electromagnetic field effects.

4. Heat released by the sinking of heavy metals in the mantle.

Typically, geothermal energy is transferred by conduction/ convection from the Earth's mantle and through the crust. There are also areas in which geothermal energy transfer is particularly high because the magma is especially close to the surface. Well-known examples of these "hot spots" include volcanoes, hot springs, and geysers. The "hot spots" may be useful for the collection of geothermal energy as described below.

Geothermal energy contributes to the heating of the Earth's crust and especially to large bodies of water. However, it is important to note that the energy conveyed to the crust from geothermal heat is only 1/20,000 of the amount received from the Sun. There are some small scale facilities designed to draw power from geothermal sources. Energy is collected by drilling wells into natural fractures in areas with high temperature ground water. The hot water or steam then flows upward and can be used to drive a turbine. However, it is unlikely that geothermal energy will ever represent a significant alternate power source. This is because it is difficult to harness and it is renewed at a very slow rate. Geothermal energy can be used to heat and cool a house, but it is difficult to harness enough geothermal energy to heat and cool a building of any size such as an office building, so at this point in time, geothermal energy would not be a source of energy for a town or city.

Skill 5.4   Analyzing the effects of human activities on the earth's natural resources

Pollutants are impurities in air and water that may be harmful to life. All forms of pollution have both local and global economic, aesthetic, and medical consequences. Air, land, and water pollution directly and indirectly affect human health. Pollution negatively influences the local and global economy by increasing medical costs, increasing pollution treatment costs (e.g. water and soil clean up), and decreasing agricultural yields. Finally, all types of pollution decrease natural beauty and diminish enjoyment of nature and the outdoors.

Air pollution, possibly the most damaging form of pollution, has both local and global consequences. Air pollution results largely from the burning of fuels. Major sources of air pollution include transportation, industrial processes, heat and power generation, and the burning of solid waste. At the local level, air pollution negatively affects quality of life by increasing medical problems and decreasing comfort and enjoyment of outdoor activities. Air pollution can cause medical problems ranging from simple throat or eye irritations to asthma and lung cancer. Globally, air pollution threatens the Earth's ozone layer and may cause global warming. Ozone depletion and global warming increase the risk of sun related health problems (e.g. skin cancer) and threaten the Earth's ecological balance and biodiversity. Possible solutions to air pollution are government controls on fuel types, industry combustion standards, and the development and use of alternative, cleaner sources of energy.

Land pollution is the destruction of the Earth's surface resulting from improper industrial and urban waste disposal, damaging agricultural practices, and mining and mineral exploitation. Land pollution greatly affects aesthetic appeal and human health. At the local level, improper waste disposal threatens the health of people living in the affected areas. Waste accumulation attracts pests and creates unsightly, dirty living conditions. Globally, damage and depletion of the soil by improper agricultural practices has great economic consequences. Overuse of pesticides and herbicides and depletion of soil nutrients causes long-term damage to the soil, leading to decreased crop yield in the future. Close regulation of waste disposal and agricultural practices is the most effective strategy for the prevention of land pollution.

Water pollution, perhaps the most prevalent form of pollution, greatly affects human health and the economy. The agricultural industry is the leading contributor to water pollution. Rain runoff carrying pesticides, herbicides, and fertilizer quickly pollutes oceans, lakes, and rivers. Also contributing to water pollution are industrial effluents, sewage, and domestic waste. Water pollution negatively affects the economy because clean-up and treatment of polluted water is very costly. Spills from ships and barges carrying large quantities of oil pollute beaches and harm fish. In addition, consumption of polluted water causes health problems and increases associated medical costs. Finally, polluted water disrupts aquatic ecosystems, decreasing the availability of fish and other aquatic resources. Like air and land pollution, limiting water pollution requires strict governmental control. Proper oversight requires the implementation and enforcement of environmental standards regulating agricultural and industrial discharge into bodies of water.

All acids contain hydrogen. Acidic substances from factories and car exhausts dissolve in rain water forming **acid rain.** Acid rain forms predominantly from pollutant oxides in the air (usually nitrogen-based NO_x or sulfur-based SO_x), which become hydrated into their acids (nitric or sulfuric acid). When the rain falls into stone, the acids can react with metallic compounds and gradually wear the stone away.

Skill 5.5   Demonstrating knowledge of methods for conserving and protecting natural resources

Stewardship is the responsible management of resources entrusted to one. Because human presence and activity has such a drastic impact on the environment, humans are the stewards of the Earth. In other words, it is the responsibility of humans to balance their needs as a population with the needs of the environment and all the Earth's living creatures and the available resources. Stewardship requires the regulation of human activity to prevent, reduce, and mitigate environmental degradation. An important aspect of stewardship is the preservation of resources and ecosystems for future generations of humans. Finally, the concept of stewardship often, but not necessarily, draws from religious, theological, or spiritual thought and principles.

Strategies for the management of renewable resources focus on balancing the immediate demand for resources with long-term sustainability. In addition, renewable resource management attempts to optimize the quality of the resources. For example, scientists may attempt to manage the amount of timber harvested from a forest, balancing the human need for wood with the future viability of the forest as a source of wood. Companies that harvest trees also plant trees, usually more than they harvest. Scientists attempt to increase timber production by fertilizing, manipulating trees genetically, and managing pests and density. Similar strategies exist for the management and optimization of water sources, air quality, and other plants and animals.

The main concerns in nonrenewable resource management are conservation, allocation, and environmental mitigation. Policy makers, corporations, and governments must determine how to use and distribute scarce and non-renewable resources. Decision makers must balance the immediate demand for resources with the need for resources in the future. This determination is often the cause of conflict and disagreement. Finally, scientists attempt to minimize and mitigate the environmental damage caused by resource extraction. Scientists devise methods of harvesting and using resources that do not unnecessarily impact the environment. After the extraction of resources from a location, scientists devise plans and methods to restore the environment to as close to its original state as possible.

0006 UNDERSTAND THE DIVERSITY OF LIVING ORGANISMS AND THEIR CLASSIFICATION

Skill 6.1   Recognizing major characteristics of organisms from different biological kingdoms

Kingdom Monera - bacteria and blue-green algae, prokaryotic, having no true nucleus, unicellular.

Bacteria are classified according to their morphology (shape). **Bacilli** are rod shaped, **cocci** are round, and **spirillia** are spiral shaped. The **gram stain** is a staining procedure used to identify bacteria. Gram positive bacteria pick up the stain and turn purple. Gram negative bacteria do not pick up the stain and are pink in color.

Kingdom Protista - eukaryotic, unicellular, some are photosynthetic, some are consumers. Microbiologists use methods of locomotion, reproduction, and how the organism obtains its food to classify protista.

Methods of locomotion - Flagellates have a flagellum, ciliates have cilia, and amoeboid move through use of pseudopodia.

Methods of reproduction - binary fission is simply dividing in half and is asexual. All new organisms are exact clones of the parent. Sexual modes provide more diversity. Bacteria can reproduce sexually through conjugation, where genetic material is exchanged.

Methods of obtaining nutrition - photosynthetic organisms or producers, convert sunlight to chemical energy, consumers or heterotrophs eat other living things. Saprophytes are consumers that live off dead or decaying material.

Kingdom Fungi - eukaryotic, multicellular, absorptive consumers, contain a chitin cell wall.

Kingdom Plantae (Plant Kingdom):
NONVASCULAR PLANTS - small in size, do not require vascular tissue (xylem and phloem) because individual cells are close to their environment. The nonvascular plants have no true leaves, stems or roots.

Division Bryophyta - mosses and liverworts, these plants have a dominant gametophyte generation. They possess rhizoids, which are root-like structures. Moisture in their environment is required for reproduction and absorption.

VASCULAR PLANTS - the development of vascular tissue enabled these plants to grow in size. Xylem and phloem allowed for the transport of water and minerals up to the top of the plant, as well as transport food manufactured in the leaves to the bottom of the plant. All vascular plants have a dominant sporophyte generation.

> **Division Lycophyta** - club mosses; these plants reproduce with spores and require water for reproduction.

> **Division Sphenophyta** - horsetails; also reproduce with spores. These plants have small, needle-like leaves and rhizoids. Require moisture for reproduction.

> **Division Pterophyta** - ferns; reproduce with spores and flagellated sperm. These plants have a true stem and need moisture for reproduction.

> **Gymnosperms** - The word means "naked seed". These were the first plants to evolve with seeds, which made them less dependent on water to assist in reproduction. Their seeds could travel by wind. Pollen from the male was also easily carried by the wind. Gymnosperms have cones that protect the seeds.

> **Division Cycadophyta** - cycads; these plants look like palms with cones.

> **Divison Ghetophyta** - desert dwellers.

> **Division Coniferophyta** - pines; these plants have needles and cones.

> **Divison Ginkgophyta** - the Ginkgo is the only member of this division.

> **Angiosperms (Division Anthophyta)** - Angiosperms are the largest group in the plant kingdom. They are the flowering plants and produce true seeds for reproduction.

Kingdom Anamalia (Animal Kingdom):

> **Annelida** - the segmented worms. The Annelida have specialized tissue. The circulatory system is more advanced in these worms and is a closed system with blood vessels. The nephridia are their excretory organs. They are hermaphrodidic and each worm fertilizes the other upon mating. They support themselves with a hydrostatic skeleton and have circular and longitudinal muscles for movement.

Mollusca - clams, octopus, soft bodied animals. These animals have a muscular foot for movement. They breathe through gills and most are able to make a shell for protection from predators. They have an open circulatory system, with sinuses bathing the body regions.

Arthropoda - insects, crustaceans and spiders; this is the largest group of the animal kingdom. Phylum arthropoda accounts for about 85% of all the animal species. Animals in the phylum arthropoda possess an exoskeleton made of chitin. They must molt to grow. Insects, for example, go through four stages of development. They begin as an egg, hatch into a larva, form a pupa, then emerge as an adult. Arthropods breathe through gills, trachae or book lungs. Movement varies, with members being able to swim, fly, and crawl. There is a division of labor among the appendages (legs, antennae, etc). This is an extremely successful phylum, with members occupying diverse habitats.

Echinodermata - sea urchins and starfish; these animals have spiny skin. Their habitat is marine. They have tube feet for locomotion and feeding.

Chordata - all animals with a notocord or a backbone. The classes in this phylum include Agnatha (jawless fish), Chondrichthyes (cartilage fish), Osteichthyes (bony fish), Amphibia (frogs and toads; gills which are replaced by lungs during development), Reptilia (snakes, lizards; the first to lay eggs with a protective covering), Aves (birds; warm-blooded with wings consisting of a particular shape and composition designed for flight), and Mammalia (warm blooded animals with body hair that bear their young alive, and possess mammary glands for milk production).

Skill 6.2   Comparing and contrasting prokaryotic and eukaryotic cells

The two types of cells are prokaryotic and eukaryotic. **Prokaryotic** cells consist only of bacteria and blue-green algae. Bacteria were most likely the first cells and date back in the fossil record to 3.5 billion years ago. The important things that put these cells in their own group are:

1.   They have no defined nucleus or nuclear membrane. The DNA and ribosomes float freely within the cell.

2.   They have a thick cell wall. This is for protection, to give shape, and to keep the cell from bursting.

3.   The cell walls contain amino sugars (glycoproteins). Penicillin works by disrupting the cell wall, which is bad for the bacteria, but will not harm the host.

4.   Some have a capsule made of polysaccharides which make the bacteria sticky.

5. Some have pili, which is a protein strand. This also allows for attachment of the bacteria and may be used for sexual reproduction (conjugation).

6. Some have flagella for movement.

Eukaryotic cells are found in protists, fungi, plants and animals. Some features of eukaryotic cells include:

1. They are usually larger than prokaryotic cells.

2. They contain many organelles, which are membrane bound areas for specific cell functions.

3. They contain a cytoskeleton which provides a protein framework for the cell.

4. They contain cytoplasm to support the organelles and contain the ions and molecules necessary for cell function.

Skill 6.3   Classifying organisms using various strategies and criteria

Taxonomy is the science of classification- the laws and principles covering the classification of objects. It is also interchangeably used with the term systematics, although the latter is more often thought of as including related disciplines like biogeography and evolutionary biology.

Aristotle in the 7th century was considered as the first taxonomist. The next taxonomist was Theophrastus who wrote, *De Plantes*, which covered medicinal plants only. **Carolus Linnaeus** (1707-1778) collated descriptive catalogues, in which plants and animals were classified, described, named, numbered and provided with means of identification.   He is termed the father of taxonomy. Linnaeus based his system on morphology (study of structure). Later on, evolutionary relationships (phylogeny) were also used to sort and group species. The modern classification system uses binomial nomenclature. This consists of a two word name for every species. The genus is the first part of the name and the species is the second part. Notice, in the levels explained below, that Homo sapiens is the scientific name for humans. Starting with the kingdom, the groups get smaller and more alike as one moves down the levels in the classification of humans:

Kingdom: Animalia, **Phylum:** Chordata, **Subphylum:** Vertebrata, **Class:** Mammalia, **Order:** Primate, **Family:** Hominidae, **Genus:** Homo, **Species:** sapiens

Species are defined by the ability to successfully reproduce with members of their own kind.

The four main considerations in taxonomy are –

Classification (hierarchical arrangement of taxa)
Descriptions (for each level)
Nomenclature (internationally accepted binomial system in Latin)
Identification aids (keys, pictures, collections etc.,)

Classification: There are two main types of classification:

1. Artificial classification – based on a few arbitrarily selected characters, e.g., healing properties of plants. Linnaeus used the number of stamens and pistils in his so called sexual system of classification. Linnaeus's system came under criticism in the 19th century.

2. Natural classification – In the 19th century, two Parisian biologists, de Jussieu and Adanson used a greater number of characteristics selected from a wide range of homologous (shared due to common ancestors) characters. At that time, those characters were based on morphology, but they are extended to include other characteristics including DNA sequences and behavior.

Descriptions: The hierarchical classification provides a structure for recording information at any level. Each taxon has both a name and description. These descriptions are unique to that particular taxon.

Nomenclature: Each taxon has one scientific name. This name is of absolutely essential function in communication and retrieval of biological information. They are in Latinized form, often with endings (-formes, -idae, -ceae). By international agreement, taxonomists who name taxa follow the codes of nomenclature: International Code of Zoological Nomenclature for animals, International Code of Botanical Nomenclature for plants and fungi, and International Code of Nomenclature for bacteria and actinomycetes.

The names of genera are single words with capital initial letters, usually underlined in hand writing and italicized in print, e.g., Drosophila, Quercus

The names of species are binomials, the first genus name and specific name or epithet, e.g., Passiflora edulis, Felix tigris.

Subdivisions blow species have trinomials, e.g., Cervus elaphus maral, Vicia faba anatolica.

The relationship between taxonomy and the history of evolution

Evolutionary taxonomy classifies groups according to the course of evolution. It states that the process of evolution produces the natural groups of natural classification. The similarities of the natural groups are due to common ancestry, which can only be inferred, not observed. It is inferred by sharing of the characters, not only due to common ancestry but also due to convergence. The evolutionary taxonomists have to differentiate between these two. They have to identify homologies (ancestral) and exclude analogies (convergent characters). The techniques of evolutionary taxonomy are imperfect. Evolutionary biologists are aware of this and in such a situation, they prefer to classify according to phenotypic divergence.

Skill 6.4   Developing and using dichotomous keys

A dichotomous key is a biological tool for identifying unknown organisms to some taxonomic level. It is constructed of a series of couplets, each consisting of two statements describing characteristics of a particular organism or group of organisms.   A choice between the statements is made that bet fits the organism in question. The statements typically begin with broad characteristics and become narrower as more choices are needed.

Example A - Numerical key

    1. Seeds round – soybeans
    1. Seeds oblong – 2
    2. Seeds white – northern beans
    2. Seeds black – black beans

Example B - Alphabetical key

        A. Seeds oblong – B
          B. Seeds white – northern beans
          B. Seeds black – black beans
        A. Seeds round - soybeans

Skill 6.5 Analyzing the evolutionary basis of modern classification systems

The current five kingdom system separates prokaryotes from eukaryotes. The prokaryotes belong to the kingdom monera while the eukaryotes belong to either kingdom protista, plantae, fungi, or animalia. Recent comparisons of nucleic acids and proteins between different groups of organisms have led to problems concerning the five kingdom system. Based on these comparisons, alternative kingdom systems have emerged. Six and eight kingdoms as well as a three domain system have been proposed as possibly more accurate classification systems. It is important to note that classification systems evolve as more information regarding characteristics and evolutionary histories of organisms arise.

Evolutionary taxonomy classifies groups according to the course of evolution. It states that the process of evolution produces the natural groups of natural classification. The similarities of the natural groups are due to common ancestry, which can only be inferred, not observed. It is inferred by sharing of the characters, not only due to common ancestry but also due to convergence. The evolutionary taxonomists have to differentiate between these two. They have to identify homologies (ancestral) and exclude analogies (convergent characters). The techniques of evolutionary taxonomy are imperfect.

0007 UNDERSTAND THE STRUCTURE AND FUNCTION OF LIVING SYSTEMS

Skill 7.1 Demonstrating knowledge of the relationship between the structure of plant and animal cell organelles and basic cell functions

The structure of the cell is often related to the cell's function. Root hair cells differ from flower stamens or leaf epidermal cells. They all have different functions.

Animal cells – The nucleus of an animal cell is a round body inside the cell. It controls the cell's activities. The nuclear membrane contains threadlike structures called chromosomes. The genes are units that control cell activities found in the nucleus. The cytoplasm has many structures in it. Vacuoles contain the food for the cell. Other vacuoles contain waste materials. Animal cells differ from plant cells because they have cell membranes rather than cell walls. Many other organelles are contained in animal cells, some of which are important for the synthesis of DNA and RNA. Other organelles inside the animal cell include lysosomes, mitochondria, endoplasmic reticulum, and the Golgi apparatus.

Plant cells – Plant cells have cell walls. A cell wall differs from a cell membrane. The cell membrane is very thin and is a part of the cell. The cell wall is thick and is a nonliving part of the cell. Chloroplasts are bundles of chlorophyll used to turn sunlight into energy. Plant cells also contain organelles such as ribosomes, mitochondria, the Golgi apparatus, lysosomes, and endoplasmic reticulum.

Single celled organism – A single celled organism is called a **protist.** When you look under a microscope the animal-like protists are called **protozoans.** They do not have chloroplasts. They are usually classified by the way they move for food. Amoebas engulf other protists by flowing around and over them. The paramecium has a hair-like structure that allows it to move back and forth like using tiny oars on a boat for searching for food. The euglena is an example of a protozoan that moves with a tail-like structure called a flagellum.

Plant-like protists have cell walls and float in the ocean. **Bacteria** are the simplest protists. A bacterial cell is surrounded by a cell wall, but there is no nucleus inside the cell. Most bacteria do not contain chlorophyll so they do not make their own food. The classification of bacteria is by shape. Cocci are round, bacilli are rod-shaped, and spirilla are spiral-shaped.

Below are the organelles found in eukaryotic cells, such as those found in plants and animals.

1. Nucleus - The brain of the cell. The nucleus contains:

> **chromosomes**- DNA, RNA and proteins tightly coiled to conserve space while providing a large surface area.

> **chromatin** - loose structure of chromosomes. Chromosomes are called chromatin when the cell is not dividing.

> **nucleoli** - where ribosomes are made. These are seen as dark spots inside the nucleus.

> **nuclear membrane** - contains pores which let RNA out of the nucleus. The nuclear membrane is continuous with the endoplasmic reticulum which allows the membrane to expand or shrink if needed.

2. Ribosomes – These are the site of protein synthesis. Ribosomes may be free floating in the cytoplasm or attached to the endoplasmic reticulum. There may be up to a half a million ribosomes in a cell, depending on how much protein is made by the cell.

3. Endoplasmic Reticulum - These are folded and provide a large surface area. They are the "roadway" of the cell and allow for transport of materials. The lumen of the endoplasmic reticulum helps to keep materials out of the cytoplasm and headed in the right direction. The endoplasmic reticulum is capable of building new membrane material. There are two types:

> **Smooth Endoplasmic Reticulum** - contain no ribosomes on their surface.

> **Rough Endoplasmic Reticulum** - contain ribosomes on their surface. This form of ER is abundant in cells that make many proteins, like in the pancreas, which produces many digestive enzymes.

4. Golgi Complex or Golgi Apparatus - This structure is stacked to increase surface area. The Golgi Complex functions to sort, modify and package molecules that are made in other parts of the cell. These molecules are either sent out of the cell or to other organelles within the cell.

5. Lysosomes – These are found mainly in animal cells. These contain digestive enzymes that break down food, substances not needed, viruses, damaged cell components, and eventually the cell itself. It is believed that lysosomes are responsible for the aging process.

6. Mitochondria – The mitochondria are large organelles that make ATP to supply energy to the cell. Muscle cells have many mitochondria because they use a great deal of energy. The folds inside the mitochondria are called cristae. They provide a large surface where the reactions of cellular respiration occur. Mitochondria have their own DNA and are capable of reproducing themselves if a greater demand is made for additional energy. Mitochondria are found primarily in animal cells.

7. Plastids – Plastids are found in photosynthetic organisms only. They are similar to the mitochondria due to their double membrane structure. They also have their own DNA and can reproduce if increased capture of sunlight becomes necessary. There are several types of plastids:

> **Chloroplasts** - green, function in photosynthesis. They are capable of trapping sunlight.

> **Chromoplasts** - make and store yellow and orange pigments; they provide color to leaves, flowers and fruits.

> **Amyloplasts** - store starch and are used as a food reserve. They are abundant in roots like potatoes.

8. Cell Wall – The cell wall is found in plant cells only. It is composed of cellulose and fibers. It is thick enough for support and protection, yet porous enough to allow water and dissolved substances to enter. Cell walls are cemented to each other.

9. Vacuoles – They hold stored food and pigments. Vacuoles are very large in plants. This allows them to fill with water in order to provide turgor pressure. Lack of turgor pressure causes a plant to wilt.

10. Cytoskeleton – It is composed of protein filaments attached to the plasma membrane and organelles. They provide a framework for the cell and aid in cell movement. They constantly change shape and move about. Three types of fibers make up the cytoskeleton:

> **Microtubules** - largest of the three; makes up cilia and flagella for locomotion. Flagella grow from a basal body. Some examples are sperm cells, and tracheal cilia. Centrioles are also composed of microtubules. They form the spindle fibers that pull the cell apart into two cells during cell division. Centrioles are not found in the cells of higher plants.

> **Intermediate Filaments** - they are smaller than microtubules but larger than microfilaments. They help the cell to keep its shape.

Microfilaments - smallest of the three, they are made of actin and small amounts of myosin (like in muscle cells). They function in cell movement such as cytoplasmic streaming, endocytosis, and ameboid movement. This structure pinches the two cells apart after cell division to form two cells.

Skill 7.2   Relating the structure and function of macromolecules in cells

All materials of life are ultimately derived from relatively simple elements and compounds. Once they are assimilated by a living organism, these materials form the building blocks of complex compounds. Many of these complex compounds are later broken down by that organism or one that ate it into simple elemental forms again. This cycle involves the elements of carbon, oxygen, hydrogen and nitrogen and a great variety of minerals including phosphorus and sulfur.

The carbon atom can enter into thousands of different combinations within one organism and then, within others as the materials are passed on. Because carbon has the ability to form four covalent bonds, it can form long chains or branched chains and it can bond to a wide variety of other atoms. The study of compounds containing carbon is called organic chemistry since all living things (animal or plant) are comprised of carbon compounds.

Macromolecules are large molecules whose mass is bonded covalently. Macromolecules commonly found in cells include lipids, DNA, RNA, proteins, and carbohydrates. DNA and RNA are covered more fully in competency 8.

A compound consists of two or more elements. There are four major chemical compounds found in the cells and bodies of living things. These include carbohydrates, lipids, proteins and nucleic acids.

Monomers ('mono' means one) are the simplest unit of structure. **Monomers** can be combined to form **polymers** ('poly' means many), or long chains, making a large variety of molecules possible. Monomers combine through the process of condensation reactions (also called dehydration synthesis reactions). In this process, one molecule of water is removed from between the adjoining molecules. In order to break the molecules apart in a polymer, water molecules are added between monomers, thus breaking the bonds between them. This is called hydrolysis.

Carbohydrates contain a ratio of two hydrogen atoms for each carbon and oxygen $(CH_2O)_n$. Carbohydrates include sugars and starches. They function in the release of energy. **Monosaccharides** are the simplest sugars and include glucose, fructose, and galactose. They are major nutrients for cells. In cellular respiration, the cells extract the energy in glucose molecules. **Disaccharides** are made by joining two monosaccharides by condensation to form a glycosidic linkage (covalent bond between two monosaccharides). Maltose is formed from the combination of two glucose molecules, lactose is formed from joining glucose and galactose, and sucrose is formed from the combination of glucose and fructose. **Polysaccharides** consist of many monomers joined. They are storage material hydrolyzed as needed to provide sugar for cells or building material for structures protecting the cell. Examples of polysaccharides include starch, glycogen, cellulose and chitin.

> **Starch** - major energy storage molecule in plants. It is a polymer consisting of glucose monomers.

> **Glycogen** - major energy storage molecule in animals. It is made up of many glucose molecules.

> **Cellulose** - found in plant cell walls, its function is structural. Many animals lack the enzymes necessary to hydrolyze cellulose, so it simply adds bulk (fiber) to the diet.

> **Chitin** - found in the exoskeleton of arthropods and fungi. Chitin contains an amino sugar (glycoprotein).

Lipids are composed of glycerol (an alcohol) and three fatty acids. Lipids are **hydrophobic** (water fearing) and will not mix with water. There are three important families of lipids: fats, phospholipids and steroids.

> **Fats** consist of glycerol (alcohol) and three fatty acids. Fatty acids are long carbon skeletons. The nonpolar carbon-hydrogen bonds in the tails of fatty acids make them hydrophobic. Fats are solids at room temperature and come from animal sources (butter, lard).

> **Phospholipids** are a vital component in cell membranes. In a phospholipid, one or two fatty acids are replaced by a phosphate group linked to a nitrogen group. They consist of a **polar** (charged) head that is hydrophilic or water loving and a **nonpolar** (uncharged) tail which is hydrophobic or water fearing. This allows the membrane to orient itself with the polar heads facing the interstitial fluid found outside the cell and the internal fluid of the cell.

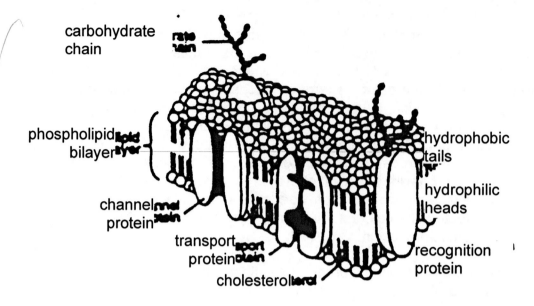

Steroids are insoluble and are composed of a carbon skeleton consisting of four inter-connected rings. An important steroid is cholesterol, which is the precursor from which other steroids are synthesized. Hormones, including cortisone, testosterone, estrogen, and progesterone, are steroids. Their insolubility keeps them from dissolving in body fluids.

Proteins compose about fifty percent of the dry weight of animals and bacteria. Proteins function in structure and aid in support (connective tissue, hair, feathers, quills), provide storage of amino acids (albumin in eggs, casein in milk), transport substances (hemoglobin), coordinate body activities through hormones (insulin), provide membrane receptor proteins, contract (muscles, cilia, flagella), defend the body (antibodies), and speed up chemical reactions (enzymes).All proteins are made of combinations of several of the twenty amino acids. An amino acid contains an amino group and an acid group. The radical group varies and defines the amino acid. Amino acids form through condensation reactions with the removal of water. The bond that is formed between two amino acids is called a peptide bond. Polymers of amino acids are called polypeptide chains. An analogy can be drawn between the twenty amino acids and the alphabet. Millions of words can be formed using an alphabet of only twenty-six letters. This diversity is also possible using only twenty amino acids. This results in the formation of many different proteins, whose structure defines the function.

There are four levels of protein structure: primary, secondary, tertiary, and quaternary.

Primary structure is the protein's unique sequence of amino acids. A slight change in primary structure can affect a protein's conformation and its ability to function. Secondary structure is the coils and folds of polypeptide chains. The coils and folds are the result of hydrogen bonds along the polypeptide backbone. The secondary structure is either in the form of an alpha helix or a pleated sheet.

The alpha helix is a coil held together by hydrogen bonds. A pleated sheet is the polypeptide chain folding back and forth. The hydrogen bonds between parallel regions hold it together. **Tertiary structure** is formed by bonding between the side chains of the amino acids. Disulfide bridges are created when two sulfhydryl groups on the amino acids bond together to form a strong covalent bond. **Quaternary structure** is the overall structure of the protein from the aggregation of two or more polypeptide chains. An example of this is hemoglobin. Hemoglobin consists of two kinds of polypeptide chains.

Nucleic acids consist of DNA (deoxyribonucleic acid) and RNA (ribonucleic acid). Nucleic acids contain the instructions for the amino acid sequence of proteins and the instructions for replicating. The monomer of nucleic acids is called a nucleotide. A nucleotide consists of a 5 carbon sugar, (deoxyribose in DNA, ribose in RNA), a phosphate group, and a nitrogenous base. The base sequence codes for the instructions. There are five bases: adenine, thymine, cytosine, guanine, and uracil. Uracil is found only in RNA and replaces the thymine. A summary of nucleic acid structure can be seen in the table below:

	SUGAR	PHOSPHATE	BASES
DNA	deoxy-ribose	present	adenine, thymine, cytosine, guanine
RNA	ribose	present	adenine, uracil, cytosine, guanine

Due to the molecular structure, adenine will always pair with thymine in DNA or uracil in RNA. Cytosine always pairs with guanine in both DNA and RNA. This allows for the symmetry of the DNA molecule seen below.

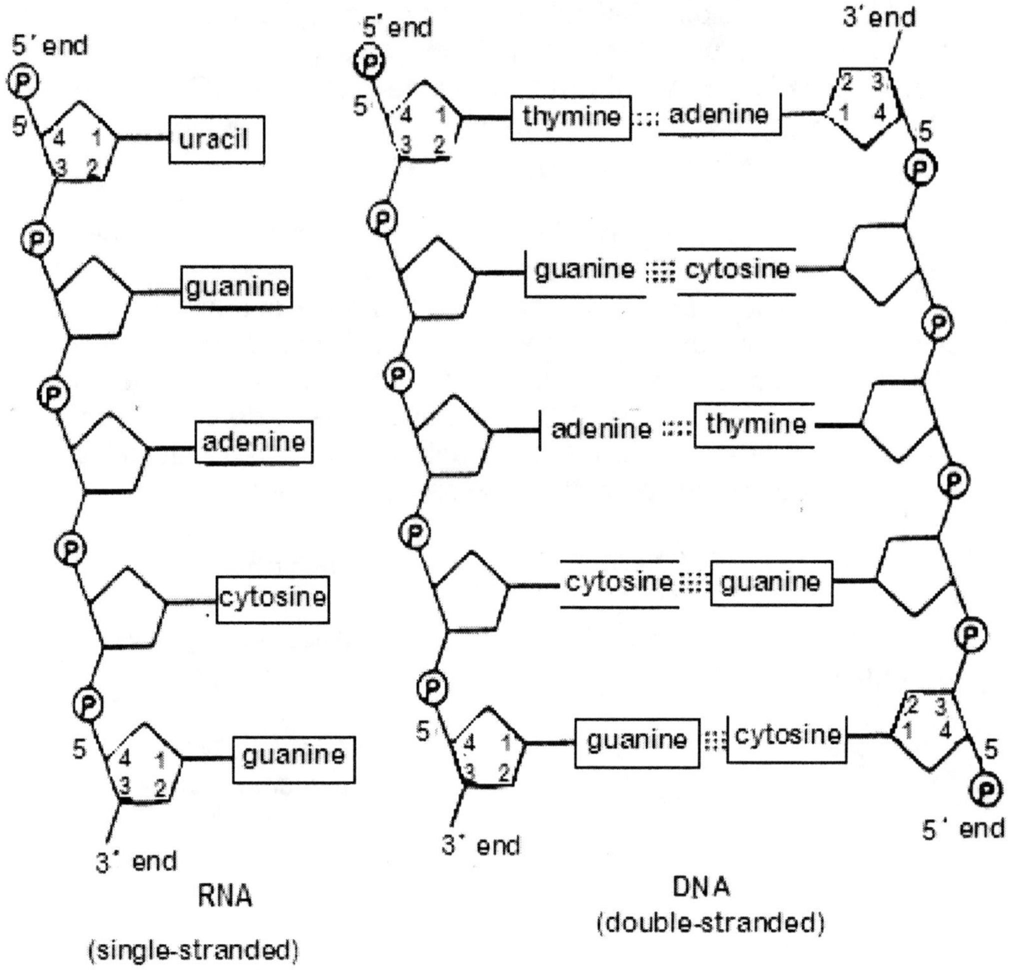

RNA
(single-stranded)

DNA
(double-stranded)

Adenine and thymine (or uracil) are linked by two covalent bonds and cytosine and guanine are linked by three covalent bonds. The guanine and cytosine bonds are harder to break apart than thymine (uracil) and adenine because of the greater number of these bonds. The DNA molecule is called a double helix due to its twisted ladder shape.

Skill 7.3   Analyzing levels of organization from the cell to the biosphere

Life is highly organized. The organization of living systems builds on levels from small to increasingly larger and more complex. All aspects, whether it is a cell or an ecosystem, have the same requirements to sustain life. Life is organized from simple to complex in the following way:

Atoms→molecules→organelles→cells→tissues→organs→organ systems→organism.

Once an organism is complete, it interacts within its environment, creating another level of organization. Two organisms together (in the case of sexual reproduction) create a **mating pair** and a mating pair reproduces to proliferate the **species**. A **population** is a group of the same species in a specific area. A **community** is a group of populations residing in the same area. Communities that are ecologically similar in regards to temperature, rainfall, and inhabitants are called **biomes**. All of the world's biomes are collectively known as the **biosphere**.

Skill 7.4   Recognizing the structure and function of cells, tissues, and organs in the major organ systems in humans

Skeletal System - The skeletal system functions in support. Vertebrates have an endoskeleton (within the flesh of the body), with muscles attached to bones. Skeletal proportions are controlled by area to volume relationships. Body size and shape is limited due to the forces of gravity. Surface area is increased to improve efficiency in all organ systems.

> The **axial skeleton** consists of the bones of the skull and vertebrae. The appendicular skeleton consists of the bones of the legs, arms and tail and shoulder girdle. Bone is a connective tissue. Parts of the bone include compact bone which gives strength, spongy bone which contains red marrow to make blood cells, yellow marrow in the center of long bones to store fat cells, and the periosteum which is the protective covering on the outside of the bone.

> A **joint** is defined as a place where two bones meet. Joints enable movement. Ligaments attach bone to bone. Tendons attach bones to muscles.

Muscular System – The function of muscles is for movement. There are three types of muscle tissue. Skeletal muscle is voluntary, so it can be moved deliberately. These muscles are attached to bones. Smooth muscle is involuntary. It is found in organs and enable functions such as digestion and respiration. Cardiac muscle is a specialized type of smooth muscle and is found in the heart. Muscles can only contract; therefore they work in antagonistic pairs to allow back and forward movement. Muscle fibers are made of groups of myofibrils, which are made of groups of sarcomeres. Actin and myosin are proteins, which make up the sarcomere.

> **Physiology of muscle contraction** - A nerve impulse strikes a muscle fiber. This causes calcium ions to flood the sarcomere. Calcium ions allow ATP to expend energy. The myosin fibers creep along the actin, causing the muscle to contract. Once the nerve impulse has passed, calcium is pumped out and the contraction ends.

Nervous System - The neuron is the basic unit of the nervous system. It consists of an axon, which carries impulses away from the cell body, the dendrite, which carries impulses toward the cell body and the cell body, which contains the nucleus. Synapses are spaces between neurons. Chemicals called neurotransmitters are found close to the synapse. The myelin sheath, composed of Schwann cells, covers the neurons and provides insulation.

Physiology of the nerve impulse - Nerve action depends on depolarization and an imbalance of electrical charges across the neuron. A polarized nerve has a positive charge outside the neuron. A depolarized nerve has a negative charge outside the neuron. Neurotransmitters turn off the sodium pump, which results in depolarization of the membrane.

This wave of depolarization (as it moves from neuron to neuron) carries an electrical impulse. This is actually a wave of opening and closing gates that allows for the flow of ions across the synapse. Nerves have an action potential. There is a threshold of the level of chemicals that must be met or exceeded in order for muscles to respond. This is called the "all or none" response.

The **reflex arc** is the simplest nerve response. The brain is bypassed. When a stimulus (like touching a hot stove) occurs, sensors in the hand send the message directly to the spinal cord. This stimulates motor neurons that contract the muscles to move the hand.

Voluntary nerve responses involve the brain. Receptor cells send the message to sensory neurons, which lead to association neurons. The message is taken to the brain. Motor neurons are stimulated and the message is transmitted to effector cells, which cause the end action.

Organization of the Nervous System - The somatic nervous system is controlled consciously. It consists of the central nervous system (brain and spinal cord) and the peripheral nervous system (nerves that extend from the spinal cord to the muscles). The autonomic nervous system is unconsciously controlled by the hypothalamus of the brain. Smooth muscles, the heart and digestion are some processes controlled by the autonomic nervous system. It works in opposition. The sympathetic nervous system works opposite of the parasympathetic nervous system. For example, if the sympathetic nervous system stimulates an action, the parasympathetic nervous system would end that action.

Neurotransmitters are chemicals released by exocytosis. Some neurotransmitters stimulate, others inhibit action.

Acetylcholine, the most common neurotransmitter; controls muscle contraction and heartbeat. The enzyme acetylcholinesterase breaks it down to end the transmission.

Epinephrine is responsible for the "fight or flight" reaction. It causes an increase in heart rate and blood flow to prepare the body for action. It is also called adrenaline.

Endorphins and enkephalins are natural pain killers and are released during serious injury and childbirth. Endorphins are also released when a person experiences intense pleasure. They are thought to be helpful in protecting a person from infection.

Digestive System - The function of the digestive system is to break food down and absorb it into the blood stream where it can be delivered to all the cells of the body for use in cellular respiration. The teeth and saliva begin digestion by breaking food down into smaller pieces and lubricating it so it can be swallowed. The lips, cheeks and tongue use the thick, mucous form of saliva to form a bolus or ball of food. Saliva contains salivary amylase which helps break down polysaccharides like starch. Swallowing forces the bolus into the esophagus. At the top end of the esophagus is the epiglottis which closes once the bolus has passed to prevent food from entering the trachea.

It is carried down the pharynx by the process of peristalsis (wave like contractions of the circular muscles and long muscles of the digestive system) and enters the stomach through the cardiac sphincter, which closes to keep food from going back up. Gastric acid in the stomach provides an optimum pH for the reactions in the stomach. In the stomach, pepsinogen and hydrochloric acid form pepsin, the enzyme that breaks down proteins. The food is broken down further by this chemical action and is churned to form chyme. Small molecules such as alcohol are absorbed directly (diffusion) from the stomach into the circulatory system.

The pyloric sphincter muscle opens to allow the food to enter the small intestine. Most nutrient absorption occurs in the small intestine by osmosis, diffusion, or active transport. It is a long tube that is folded over and over to fit the abdominal cavity. Its large surface area, accomplished by its length and protrusions into its interior called villi (which are covered with hair-like structures called microvilli) allow for a great absorptive surface into the bloodstream. Chyme is neutralized after coming from the acidic stomach to allow the enzymes found there to function. Bile emulsifies fats and neutralizes chime. Maltase, lactase, and sucrose process sugars.

Any food left after the trip through the small intestine enters the large intestine. The large intestine starts at the cecum (the junction with the small intestine), then goes through the four parts of the colon (ascending colon, transverse colon, descending colon, and rectum), then through the anal sphincter and out the anus. The large intestine functions to reabsorb water and produce vitamin K. The feces, or remaining waste, are passed out through the anus.

Carbohydrate digestion needs to be finished by using insulin secreted by the liver so that the glucose can be transformed to glycogen which is stored in the liver, adipose tissue, and muscle cells for later use for energy. When needed, the pancreas will secrete glucagon to convert the glycogen to glucose. Fat in the small intestine produces hormones which stimulate the release of lipase from the pancreas and bile from the gallbladder. Lipase breaks down fat into monoglycerides and fatty acids and bile emulsifies the fatty acids for absorption.

> **Accessory organs** - although not part of the digestive tract, these organs function in the production of necessary enzymes and bile. The pancreas makes many enzymes to break down food in the small intestine. The liver makes bile which breaks down and emulsifies fatty acids.

Respiratory System - This system functions in the gas exchange of needed oxygen and carbon dioxide waste. It delivers oxygen to the bloodstream and picks up carbon dioxide for release out of the body. Air enters the mouth and nose, where it is warmed, moistened and filtered of dust and particles. Cilia in the trachea trap unwanted material in mucus, which can be expelled. The trachea splits into two bronchial tubes and the bronchial tubes divide into smaller and smaller bronchioles in the lungs. The internal surface of the lung is composed of alveoli, which are thin-walled air sacs. These allow for a large surface area for gas exchange. The alveoli are lined with capillaries. Oxygen diffuses into the bloodstream and carbon dioxide diffuses out to be exhaled out of the lungs. The oxygenated blood is carried to the heart and delivered to all parts of the body.

The thoracic cavity holds the lungs. A muscle, the diaphragm, below the lungs is an adaptation that makes inhalation possible. As the volume of the thoracic cavity increases, the diaphragm muscle flattens out and inhalation occurs. When the diaphragm relaxes, exhalation occurs.

Circulatory System

The function of the circulatory system is to carry oxygenated blood and nutrients to all cells of the body and return carbon dioxide waste to be expelled from the lungs. In short, unoxygenated blood enters the heart through the inferior vena cava and superior vena cava. The first chamber it encounters is the right atrium. It goes through the tricuspid valve to the right ventricle to the pulmonary arteries and then to the lungs where it is oxygenated. It returns to the heart through the pulmonary vein into the left atrium. It travels through the bicuspid valve to the left ventricle where it is pumped to all parts of the body through the aorta.

Sinoatrial node (SA node) - the pacemaker of the heart. Located on the right atrium, it is responsible for contraction of the right and left atrium.

Atrioventricular node (AV node) - located on the left ventricle, it is responsible for contraction of the ventricles.

Blood vessels include:

arteries - lead away from the heart. All arteries carry oxygenated blood except the pulmonary artery going to the lungs. Arteries are under high pressure.

arterioles - arteries branch off to form smaller arterioles.

capillaries - arterioles branch off to form tiny capillaries that reach every cell. Blood moves slowest here due to the small size; only one red blood cell may pass at a time to allow for diffusion of gases into and out of cells. Nutrients are also absorbed by the cells from the capillaries.

venules - capillaries combine to form larger venules. The vessels are now carrying waste products from the cells.

veins - venules combine to form larger veins, leading back to the heart. Veins and venules have thinner walls than arteries because they are not under as much pressure. Veins contain valves to prevent the backward flow of blood due to gravity.

Components of the blood include:

>**plasma** – 60% of the blood is plasma. It contains salts called electrolytes, nutrients and waste. It is the liquid part of blood.

>**erythrocytes** - also called red blood cells; they contain hemoglobin which carries oxygen molecules.

>**leukocytes** - also called white blood cells. White blood cells are larger than red cells. They are phagocytic and can engulf invaders. White blood cells are not confined to the blood vessels and can enter the interstitial fluid between cells.

>**platelets** - assist in blood clotting. Platelets are made in the bone marrow.

>**Blood clotting factors** - the neurotransmitter that initiates blood vessel constriction following an injury is called serotonin. A material called prothrombin is converted to thrombin with the help of thromboplastin. The thrombin is then used to convert fibrinogen to fibrin, which traps red blood cells to form a scab and stop blood flow.

Lymphatic System (Immune System)

The immune system is responsible for defending the body against foreign invaders. There are two defense mechanisms: nonspecific and specific.

Nonspecific defense mechanisms – They do not target specific pathogens, but are a whole body response. The first line of defense is the body's physical barriers. The skin and mucous membranes prevent the penetration of bacteria and viruses. The second line of defense is the inflammatory response. The results are often seen as symptoms of an infection. A phagocyte is a cell which uses chemotaxis to attract, adhere to, engulf and ingest foreign bodies. Macrophages, eosinophils, and neutrophils are three types of phagocytes which ingest foreign particles.

Monocytes, made in the bone marrow and released into the bloodstream, mature to become macrophages which are the largest phagocytic cells. These cells are long-lived and depend on mitochondria for energy. In some cases these cells will display proteins from previously destroyed cells on their surfaces to warn other invading cells of their fate. These cells are referred to as antigen presenting cells (APCs).

Natural killer cells (a form of eosinophils) destroy the body's own infected cells instead of the invading the microbe directly. These cells are present in blood and lymph as large granular lymphocytes. Neutrophils make up about seventy percent of all white blood cells (leukocytes). They are phagocytes that have no mitochondria but get their energy from stored glycogen. They are short-lived, but precede the macrophages to the scene of intrusion.

The blood supply to the injured area is increased, causing redness and heat. Swelling also typically occurs with inflammation. Histamine is released by basophils and mast cells when the cells are injured. This triggers the inflammatory response. Non-fixed macrophages are attracted by the histamine. They can squeeze through the walls of capillaries to go to the area and destroy dead tissue or pathogens.

Fever is a result of an increase of white blood cells. Pyrogens are released by white blood cells, which set the body's thermostat to a higher temperature. This inhibits the growth of microorganisms. It also increases metabolism to increase phagocytosis and body repair.

Specific defense mechanisms - They recognize foreign material and responds by destroying the invader. These mechanisms are specific and diverse. They are able to recognize individual pathogens. They also have recognition of foreign material versus the self. Memory of the invaders provides immunity upon further exposure.

> **antigen** - any foreign particle that invades the body.

> **antibody** - manufactured by the body, they recognize and latch onto antigens, hopefully destroying them. They recognize 'self' versus 'non-self' and they have memory so are specific once exposed to specific antigen (providing immunity).

> **immunity** - this is the body's ability to recognize and destroy an antigen before it causes harm. Active immunity develops after recovery from an infectious disease (chicken pox) or after a vaccination (mumps, measles, rubella). Passive immunity may be passed from one individual to another. It is not permanent. A good example is the immunities passed from mother to nursing child.

Excretory System

The main function of the excretory system is to rid the body of toxins and wastes. Excretion of unwanted gases like carbon dioxide is the function of the circulatory and respiratory systems. The sweat glands of the skin also provide a means of excretion. The liver regulates glycogen storage, plasma protein synthesis and drug detoxification. It also changes toxic ammonia to urea and sends it to the kidneys.

The function of the kidneys is to rid the body of nitrogenous wastes in the form of urea. The kidneys (2) are found on either side of the spinal column near the small of the back. They are bean-shaped organs responsible for removing wastes from the blood and keeping blood pressure normalized. They have an outer layer called the cortex and an inner core called the medulla. The functional unit of excretion is the nephron, many of which make up the kidneys. The "tops' of the nephrons are in the cortex while the long tubule part is in the medulla. Nephrons have lots of blood supply from the glomerulus. Antidiuretic hormone (ADH) which is made in the hypothalamus gland and stored in the pituitary is released when differences in osmotic balance occur. This will cause more water to be reabsorbed back into the blood. As the blood becomes more dilute, ADH release ends. ADH is regulated by a negative feedback loop.

The Bowman's capsule contains the glomerulus, a tightly packed group of capillaries. The glomerulus is under high pressure. Waste and fluids leak out due to pressure. Filtration is not selective in this area. Selective secretion by active and passive transport occur in the proximal convoluted tubule. Unwanted molecules are secreted into the filtrate. Selective secretion also occurs in the loop of Henle. Salt is actively pumped out of the tube and much water is lost due to the hyperosmosity of the inner part (medulla) of the kidney. As the fluid enters the distal convoluted tubule, more water is reabsorbed. Urine forms in the collecting duct, which leads to the ureter then to the bladder where it is stored. Urine is passed from the bladder through the urethra. The amount of water reabsorbed back into the body is dependent upon how much water or fluids an individual has consumed. Urine can be very dilute or very concentrated if dehydration is present.

Endocrine System

The function of the endocrine system is to manufacture proteins called hormones. Hormones are released into the bloodstream and are carried to a target tissue where they stimulate an action. Hormones may build up over time to cause their effect, as in puberty or the menstrual cycle.

Hormone activation - Hormones are specific and fit receptors on the target tissue cell surface. The receptor activates an enzyme which converts ATP to cyclic AMP. Cyclic AMP (cAMP) is a second messenger from the cell membrane to the nucleus. The genes found in the nucleus turn on or off to cause a specific response.

There are two classes of hormones. **Steroid hormones** come from cholesterol. Steroid hormones cause sexual characteristics and mating behavior. Hormones include estrogen and progesterone in females and testosterone in males.

Peptide hormones are made in the hypothalamus, pituitary gland (anterior and posterior), adrenal glands on the kidneys, pineal body, thyroid gland and the pancreas. They include the following:

> **Hypothalamus:**
>
> > **Thyrotropin-releasing hormone (TRH)** – release TSH from anterior pituitary, stimulate PRL release
> >
> > **Gonadotropin-releasing hormone (GnRH)** – Release of FSH and LH from anterior pituitary
> >
> > **Growth hormone-releasing hormone (GHRH)** – release GH from anterior pituitary
> >
> > **Corticotropin-releasing hormone (CRH)** – release ACTH from anterior pituitary
> >
> > **Vasopressin** – increases water reabsorption in the kidney by stimulating release of ADH
> >
> > **Somatostatin (also known as growth hormone-inhibiting hormone) (SS or GHIH)** – inhibit release of GH and TSH
> >
> > **Prolactin inhibiting hormone or Dopamine (PIH or DA)** – inhibit release of prolactin and TSH
>
> **Anterier Pituitary:**
>
> > **Follicle stimulating hormone (FSH)** - production of sperm or egg cells
> >
> > **Luteinizing hormone (LH)** - functions in ovulation in female and stimulates production of testosterone in male
> >
> > **Growth hormone (GH)** - stimulates growth

Prolactin (PRL) – stimulates milk production and sexual gratification (also known as Luteotropic hormone)

Adrenocorticotropic hormone (ACTH) – responsible for synthesis of corticosteroids in adrenocortical cells

Thyroid-stimulating hormone (TSH) – stimulates the thyroid to secrete thyroxin and triiodothyromine

Posterior Pituitary:

Antidiuretic hormone (ADH) - assists in retention of water

Oxytocin - stimulates labor contractions at birth and let-down of milk

Pineal body:

Melatonin - regulates circadian rhythms and seasonal changes, antioxidant, causes drowsiness

Pancreas:

Insulin - decreases glucose level in blood

Glucagon - increases glucose level in blood

Somatostatin – inhibit release of insulin, inhibit release of glucagon

Adrenal glands:

Epinephrine (adrenalin) - causes fight or flight reaction of the nervous system

Norepinephrine (noradrenalin) – involved in fight or flight reaction

Dopamine – increase heart rate and blood pressure

Thyroid glands:

Thyroxin - increases metabolic rate

Calcitonin - removes calcium from the blood

Although you probably won't be tested on individual hormones, be aware that hormones work on a feedback system. The increase or decrease in one hormone may cause the increase or decrease in another. Releasing hormones cause the release of specific other hormones or, in a negative feedback loop, cause another hormone to cease to be released.

Reproductive System

Sexual reproduction greatly increases diversity due to the many combinations possible through meiosis and fertilization. Gametogenesis is the production of the sperm and egg cells. Spermatogenesis begins at puberty in the male. One spermatozoa produces four sperm. The sperm mature in the seminiferous tubules located in the testes. Oogenesis, the production of egg cells is usually complete by the birth of a female. Egg cells are not released until menstruation begins at puberty. Meiosis forms one ovum with all the cytoplasm and three polar bodies which are reabsorbed by the body. The ovum are stored in the ovaries and released each month from puberty to menopause.

Path of the sperm - sperm are stored in the seminiferous tubules in the testes where they mature. Mature sperm are found in the epididymis located on top of the testes. After ejaculation, the sperm travels up the vas deferens where they mix with semen made in the prostate and seminal vesicles and travel out the urethra.

Path of the egg - eggs are stored in the ovaries. Ovulation releases the egg into the fallopian tubes which are ciliated to move the egg along. Fertilization normally occurs in the fallopian tube. If fertilization resulting in pregnancy does not occur, the egg passes through the uterus and is expelled through the vagina during menstruation. Levels of progesterone and estrogen stimulate menstruation and are affected by the implantation of a fertilized egg so menstruation does not occur.

Pregnancy - if fertilization occurs, the fertilized egg begins to divide, becoming a zygote which implants in about two to three days in the uterus. Implantation promotes secretion of human chorionic gonadotrophin (HCG)which is detected in pregnancy tests. The HCG keeps the level of progesterone elevated to maintain the uterine lining in order to feed the developing embryo until the umbilical cord forms. Labor is initiated by oxytocin, which causes labor contractions and dilation of the cervix. Prolactin and oxytocin cause the production of milk.

Skill 7.5   Comparing how various organisms carry out basic life processes (e.g., reproducing, obtaining nutrients, maintaining homeostasis)

Members of the five different kingdoms of the classification system of living organisms often differ in their basic life functions. Here we compare and analyze how members of the five kingdoms obtain nutrients, excrete waste, and reproduce.

Bacteria are prokaryotic, single-celled organisms that lack cell nuclei. The different types of bacteria obtain nutrients in a variety of ways. Most bacteria absorb nutrients from the environment through small channels in their cell walls and membranes (chemotrophs) while some perform photosynthesis (phototrophs). Chemoorganotrophs use organic compounds as energy sources while chemolithotrophs can use inorganic chemicals as energy sources. Depending on the type of metabolism and energy source, bacteria release a variety of waste products (e.g. alcohols, acids, carbon dioxide) to the environment through diffusion.

All bacteria reproduce through binary fission (asexual reproduction) producing two identical cells. Bacteria reproduce very rapidly, dividing or doubling every twenty minutes in optimal conditions. Asexual reproduction does not allow for genetic variation, but bacteria achieve genetic variety by absorbing DNA from ruptured cells and conjugating or swapping chromosomal or plasmid DNA with other cells.

Animals are multicellular, eukaryotic organisms. All animals obtain nutrients by eating food (ingestion). Some types of animals derive nutrients from eating plants only (herbivores) while other animals eat only animals (carnivores), and yet others like humans eat both plants and animals (omnivores). Animal cells perform respiration that converts food molecules, mainly carbohydrates and fats, into energy. The excretory systems of animals, like animals themselves, vary in complexity. Simple invertebrates eliminate waste through a single tube, while complex vertebrates have a specialized system of organs that process and excrete waste.

Most animals, unlike bacteria, exist in two distinct sexes. Members of the female sex give birth or lay eggs. Some less developed animals can reproduce asexually. For example, flatworms can divide in two and some unfertilized insect eggs can develop into viable organisms. Most animals reproduce sexually through various mechanisms. For example, aquatic animals reproduce by external fertilization of eggs, while mammals reproduce by internal fertilization. More developed animals possess specialized reproductive systems and cycles that facilitate reproduction and promote genetic variation.

Plants, like animals, are multi-cellular, eukaryotic organisms. Plants obtain nutrients from the soil through their root systems and convert sunlight into energy through photosynthesis. Many plants store waste products in vacuoles or organs (e.g. leaves, bark) that are discarded. Some plants also excrete waste through their roots.

More than half of the plant species reproduce by producing seeds from which new plants grow. Depending on the type of plant, flowers or cones produce seeds. Other plants reproduce by spores, tubers, bulbs, buds, and grafts. The flowers of flowering plants contain the reproductive organs. Pollination is the joining of male and female gametes that is often facilitated by movement by wind or animals.

Fungi are eukaryotic, mostly multi-cellular organisms. All fungi are heterotrophs, obtaining nutrients from other organisms. More specifically, most fungi obtain nutrients by digesting and absorbing nutrients from dead organisms. Fungi secrete enzymes outside of their body to digest organic material and then absorb the nutrients through their cell walls.

Most fungi can reproduce asexually and sexually. Different types of fungi reproduce asexually by mitosis, budding, sporification, or fragmentation. Sexual reproduction of fungi is different from sexual reproduction of animals. The two mating types of fungi are plus and minus, not male and female. The fusion of hyphae, the specialized reproductive structure in fungi, between plus and minus types produces and scatters diverse spores.

Protists are eukaryotic, single-celled organisms. Most protists are heterotrophic, obtaining nutrients by ingesting small molecules and cells and digesting them in vacuoles. All protists reproduce asexually by either binary or multiple fission. Like bacteria, protists achieve genetic variation by exchange of DNA through conjugation.

Skill 7.6   Explaining the major chemical processes (e.g., photosynthesis, cellular respiration, transport mechanisms) that support living organisms

Some of the major chemical processes in plants include:

Photosynthesis is the process by which plants make carbohydrates from the energy of the sun, carbon dioxide, and water. Oxygen is a waste product. Photosynthesis occurs in the chloroplast where the pigment chlorophyll traps sun energy. The summary equation for photosynthesis is:

$$6\,CO_2 + 6\,H_2O + light + chlorophyll \rightarrow C_6H_{12}O_6 \text{ (glucose)} + O_2$$

It is divided into two major steps:

> **Light Reactions** - Sunlight is trapped, water is split, and oxygen is given off. ATP is made and hydrogens reduce NADP to $NADPH_2$. The light reactions occur in light. The products of the light reactions enter into the dark reactions (Calvin cycle).

> **Dark Reactions** - Carbon dioxide enters during the dark reactions which can occur with or without the presence of light. The energy transferred from $NADPH_2$ and ATP allow for the fixation of carbon into glucose.

Respiration - during times of decreased light, plants break down the products of photosynthesis through cellular respiration. Glucose, with the help of oxygen, breaks down and produces carbon dioxide and water as waste. Approximately fifty percent of the products of photosynthesis are used by the plant for energy.

Transpiration - water travels up the xylem of the plant through the process of transpiration. Water sticks to itself (cohesion) and to the walls of the xylem (adhesion). As it evaporates through the stomata of the leaves, the water is pulled up the column from the roots. Environmental factors such as heat and wind increase the rate of transpiration. High humidity will decrease the rate of transpiration.

Active transport of food – see Active Transport below

Endocytosis – Endocytosis is the process that brings materials into cells by enclosing them in a sphere of cell membrane, then pinching off a vesicle within the cytoplasm. An important example is the entry of nitrogen-fixing bacteria into legume root cells. Many protists use phagotrophy, particle-feeding, to ingest food such as bacteria and other small particles.

Exocytosis – Exocytosis is the transport of materials out of the cells by enclosing them in a sphere of membrane known as a vesicle. Vesicles easily fuse to the cell membrane and then the side facing outside the cell opens and the material is dumped.

In animals, chemical processes include:

Animal respiration takes in oxygen and gives off waste gases. For instance a fish uses its gills to extract oxygen from the water. Bubbles are evidence that waste gasses are expelled. Respiration without oxygen is called anaerobic respiration. Anaerobic respiration in animal cells is also called lactic acid fermentation. The end product is lactic acid.

Cellular respiration: **Mitochondria** are large organelles that are the site of cellular respiration, the production of ATP that supplies energy to the cell.

Muscle cells have many mitochondria because they use a great deal of energy. Mitochondria have their own DNA, RNA, and ribosomes and are capable of reproducing by binary fission if there is a great demand for additional energy. Mitochondria have two membranes: a smooth outer membrane and a folded inner membrane. The outer membrane contains many large transport proteins to allow for large molecules to enter. It also includes proteins for converting lipid substrates into forms to be used by the matrix. The folds inside the mitochondrial inner membrane are called cristae. They provide a large surface area for cellular respiration to occur. Oxidation phosphorylation takes place here. Three major proteins are found here: (1) the proteins involved in the oxidation reactions of the respiratory chain, (2) the enzyme complex ATP synthetase for making ATP, and (3) the transport proteins for regulating the transfer into and out of the matrix. The matrix is the site of the Kreb Cycle and it contains copies of the mitochondrial DNA genome, specialized ribosomes, tRNAs, and various enzymes.

There are many chemical and biochemical reactions within the animal body. All the steps to digestion involve chemical reactions. Hormones are chemicals, so any place a hormone is involved there will be one or more chemical reactions or processes. The exchange of oxygen for carbon dioxide in the lungs involves a chemical process. See each of the systems in the body in Skill 7.4 for an idea of some of these processes. Essentially animals are giant test tubes that are compartmentalized and organized. Opposing reactions are kept separate from each other. Regulation and homeostasis are important factors.

There are places in the bodies of animals (such as the kidney) as well as in plants where osmosis takes place. **Osmosis** is the movement of water across the cell membrane according to the relative concentration of dissolved substances in the watery solutions on the insides and outsides of cells. Dissolved materials, known as solutes, include salts (ions), sugars, and other low-molecular-weight molecules that do not pass easily through cell membranes. Osmosis occurs because water moves from an area of high concentration to an area of low concentration. If water is in high concentration, it means that the dissolved substances are in low concentration, so the water osmoses from an area of low concentration of dissolved substances to an area of high concentration of dissolved substances across a cell membrane.

If the concentration of solutes inside and outside a cell is the same, the cell is osmotically balanced so its net water content does not change. The watery medium around the cell is isotonic. However, if the solute concentration is lower outside a cell than inside, the outside solution is hypotonic to the cell. More water will enter the cell than can leave, possibly causing it to burst. Many plant and fungi cells have cell walls to prevent this. If the concentration of solutes outside the cell is higher than inside the cell, water will leave the cell, causing the cytoplasm to shrink – a process known as plasmolysis. In this case, the outside solution is described as hypertonic to the cell.

Diffusion is the movement of molecules (other than water) from a region of higher concentration to a region of lower concentration. It is the mechanism of movement of oxygen, nutrients and other molecules across the capillary walls and the movement of other molecules across membranes. Diffusion takes place through the lipid bi-layer. Different types of diffusion take place in different organs. Diffusion of ions through protein channels, diffusion of ions through ion channels, mediated transport, and facilitated diffusion are some of the types of diffusion in animals' bodies.

Spontaneous ion diffusion occurs when random motion leads particles to increase entropy by equalizing concentrations. Particles tend to move into places of lower concentration. Therefore, a concentration gradient is required, but no proteins, outside energy or flagellae are required.

Active transport requires energy in order to move a solute against the gradient. Active transport allows a cell to get needed materials while preventing other materials from entering and it helps the cell get rid of wastes. All cells possess a cell membrane that has embedded proteins. It is selectively permeable; it allows for free passage of some materials such as water, oxygen, carbon dioxide, and nitrogen but not larger molecules. The larger molecules the cell needs can pass through with the help of transporter proteins. Other transporter proteins escort wastes out of the cells. The transporter proteins are specific for the molecules they transport and fit only those molecules.

Animal digestion – some animals only eat meat (carnivores) while others only eat plants (herbivores). Many animals do both (omnivores). Nature has created animals with structural adaptations so they may obtain food through sharp teeth or long facial structures. Digestion's purpose is to break down carbohydrates, fats, and proteins. Many organs are needed to digest food. The process begins with the mouth. Certain animals, such as birds, have beaks to puncture wood or allow for large fish to be consumed. The tooth structure of a beaver is designed to cut down trees. Tigers are known for their sharp teeth used to rip hides from their prey. Enzymes are catalysts that help speed up chemical reactions by lowering effective activation energy. Enzyme rate is affected by temperature, pH, and the amount of substrate. Saliva is an enzyme that changes starches into sugars.

0008 UNDERSTAND THE PRINCIPLES AND PROCESSES OF THE INHERITANCE OF BIOLOGICAL TRAITS

Skill 8.1   Identifying the structure and function of DNA, RNA, genes, and chromosomes and analyzing their roles in storing and transmitting information

DNA and DNA REPLICATION

The modern definition of a gene is a unit of genetic information. DNA makes up genes, which in turn make up the chromosomes. DNA is wound tightly around proteins in order to conserve space. The DNA/protein combination makes up the chromosome. DNA controls the synthesis of proteins, thereby controlling the total cell activity. DNA is capable of making copies of itself.

Review of DNA structure:

1. Made of nucleotides; a five carbon sugar, phosphate group and nitrogen base (which could be adenine, guanine, cytosine or thymine).

2. Consists of a sugar/phosphate backbone which is covalently bonded. The bases are joined down the center of the molecule and are attached by hydrogen bonds which are easily broken during replication.

3. The amount of adenine equals the amount of thymine and the amount of cytosine equals the amount of guanine.

4. The shape is that of a twisted ladder called a double helix. The sugar/phosphates make up the sides of the ladder and the base pairs make up the rungs of the ladder.

DNA Replication

Enzymes control each step of the replication of DNA. The molecule untwists. The hydrogen bonds between the bases break and serve as a pattern for replication. Free nucleotides found inside the nucleus join on to form a new strand. Two new pieces of DNA are formed which are identical. This is a very accurate process. There is only one mistake for every billion nucleotides added. This is because there are enzymes (polymerases) present that proofread the molecule. In eukaryotes, replication occurs in many places along the DNA at once. The molecule may open up at many places like a broken zipper. In prokaryotic circular plasmids, replication begins at a point on the plasmid and goes in both directions until it meets itself.

Base pairing rules are important in determining a new strand of DNA sequence. For example say our original strand of DNA had the sequence as follows:

1. A T C G G C A A T A G C   This may be called our sense strand as it contains a sequence that makes sense or codes for something. The complementary strand (or other side of the ladder) would follow base pairing rules (A bonds with T and C bonds with G) and would read:
2. T A G C C G T T A T C G   When the molecule opens up and nucleotides join on, the base pairing rules create two new identical strands of DNA

1. A T C G G C A A T A G C   and   A T C G G C A A T A G C
   T A G C C G T T A T C G   2.T A G C C G T T A T C G

Protein Synthesis

It is necessary for cells to manufacture new proteins for growth and repair of the organism. Protein Synthesis is the process that allows the DNA code to be read and carried out of the nucleus into the cytoplasm in the form of RNA. This is where the ribosomes are found, which are the sites of protein synthesis. The protein is then assembled according to the instructions on the DNA. There are several types of RNA.

Messenger RNA - (mRNA) copies the code from DNA in the nucleus and takes it to the ribosomes in the cytoplasm.

Transfer RNA - (tRNA) free floating in the cytoplasm. Its job is to carry and position amino acids for assembly on the ribosome.

Ribosomal RNA - (rRNA) found in the ribosomes. They make a place for the proteins to be made. rRNA is believed to have many important functions, so much research is currently being done in this area.

Along with enzymes and amino acids, the RNA's function is to assist in the building of proteins. There are two stages of protein synthesis:

Transcription - this phase allows for the assembly of mRNA and occurs in the nucleus where the DNA is found. The DNA splits open and the mRNA reads the code and "transcribes" the sequence onto a single strand of mRNA. For example, if the code on the DNA is T A C C T C G T A C G A , the mRNA will make a complementary strand reading: A U G G A G C A U G C U (Remember that uracil replaces thymine in RNA.) Each group of three bases is called a **codon**. The codon will eventually code for a specific amino acid to be carried to the ribosome. "Start" codons begin the building of the protein and "stop" codons end transcription. When the stop codon is reached, the mRNA separates from the DNA and leaves the nucleus for the cytoplasm.

Translation - this is the assembly of the amino acids to build the protein and occurs in the cytoplasm. The nucleotide sequence is translated to choose the correct amino acid sequence. As the rRNA translates the code at the ribosome, tRNA's which contain an **anticodon** seek out the correct amino acid and bring it back to the ribosome. For example, using the codon sequence from the example above:

the mRNA reads A U G / G A G / C A U / G C U
the anticodons are U A C / C U C / G U A / C G A
the amino acid sequence would be: Methionine (start) - Glu - His - Ala.

 *Be sure to note if the table you are given is written according to the codon sequence or the anticodon sequence. It will be specified.

This whole process is accomplished through the assistance of **activating enzymes**. Each of the twenty amino acids has its own enzyme. The enzyme binds the amino acid to the tRNA. When the amino acids get close to each other on the ribosome, they bond together using peptide bonds. The start and stop codons are called nonsense codons. There is one start codon (AUG) and three stop codons. (UAA, UGA and UAG). Addition mutations will cause the whole code to shift, thereby producing the wrong protein or, at times, no protein at all.

Skill 8.2   Comparing and contrasting sexual and asexual reproduction in organisms

Bacteria reproduce by **binary fission**. This asexual process is simply dividing the bacterium in half. All new organisms are exact clones of the parent.

The purpose of cell division is to provide growth and repair in body (somatic) cells and to replenish or to create sex cells for reproduction (in sexually reproducing organisms). There are two forms of cell division in higher organisms. **Mitosis** is the division of somatic cells and **meiosis** is the division of sex cells (eggs and sperm).

The obvious advantage of asexual reproduction is that it does not require a partner. This is a huge advantage for organisms, such as the hydra, which do not move around. Not having to move around to reproduce also allows organisms to conserve energy. Asexual reproduction also tends to be faster. There are disadvantages, as in the case of regeneration, in plants if the plant is not in good condition or in the case of spore-producing plants, if the surrounding conditions are not suitable for the spores to grow. As asexual reproduction produces only exact copies of the parent organism, it does not allow for genetic variation, which means that mutations, or weaker qualities, will always be passed on. This can also be detrimental to a species that is well adapted to a particular environment when the conditions of that environment change suddenly. On the whole, asexual reproduction is more reliable because it requires fewer steps and less can go wrong.

Sexual reproduction shares genetic information between gametes, thereby producing variety in the species. This can result in a better species with an improved chance of survival. There is the disadvantage that sexual reproduction requires a partner, which in turn with many organisms requires courtship, finding a mate, and mating. Another disadvantage is that sexually reproductive organisms require special mechanisms.

Skill 8.3   Identifying processes that contribute to genetic variability (e.g., meiosis, crossing-over, mutations)

The purpose of cell division is to provide growth and repair in body (somatic) cells and to replenish or create sex cells for reproduction. There are two forms of cell division. Mitosis is the division of somatic cells and **meiosis** is the division of sex cells (eggs and sperm). The table below summarizes the major differences between the two processes.

MITOSIS	MEIOSIS
1. Division of somatic cell	1. Division of sex cells
2. Two cells result from each division	2. Four cells or polar bodies result from each division
3. Chromosome number is identical to parent cells.	3. Chromosome number is half the number of parent cells
4. For cell growth and repair	4. Recombinations provide genetic diversity

Some terms to know:

gamete - sex cell or germ cell; eggs and sperm.
chromatin - loose chromosomes; this state is found when the cell is not dividing.
chromosome - tightly coiled, visible chromatin; this state is found when the cell is dividing.
homologues - chromosomes that contain the same information. They are of the same length and contain the same genes.
diploid - 2n number; diploid chromosomes are a pair of chromosomes (somatic cells).
haploid - 1n number; haploid chromosomes are a half of a pair (sex cells).

MITOSIS

The cell cycle is the life cycle of the cell. It is divided into two stages; **Interphase** and the **mitotic division** where the cell is actively dividing. Interphase is divided into three steps; G1 (growth) period, where the cell is growing and metabolizing, S period (synthesis) where new DNA and enzymes are being made and the G2 phase (growth) where new proteins and organelles are being made to prepare for cell division. The mitotic stage consists of the stages of mitosis and the division of the cytoplasm.

The stages of mitosis and their events are as follows. Be sure to know the correct order of steps. (IPMAT)

1. Interphase - chromatin is loose, chromosomes are replicated, cell metabolism is occurring. Interphase is technically <u>not</u> a stage of mitosis.

2. Prophase - once the cell enters prophase, it proceeds through the following steps continuously, with no stopping. The chromatin condenses to become visible chromosomes. The nucleolus disappears and the nuclear membrane breaks apart. Mitotic spindles form which will eventually pull the chromosomes apart. They are composed of microtubules. The cytoskeleton breaks down and the spindles are pushed to the poles or opposite ends of the cell by the action of centrioles.

3. Metaphase - kinetechore fibers attach to the chromosomes which causes the chromosomes to line up in the center of the cell (think **m**iddle for **m**etaphase)

4. Anaphase - centromeres split in half and homologous chromosomes separate. The chromosomes are pulled to the poles of the cell, with identical sets at either end.

5. Telophase - two nuclei for with a full set of DNA that is identical to the parent cell. The nucleoli become visible and the nuclear membrane reassembles. A cell plate is visible in plant cells, whereas a cleavage furrow is formed in animal cells. The cell is pinched into two cells. Cytokinesis, or division, of the cytoplasm and organelles occurs.

Meiosis contains the same five stages as mitosis, but is repeated in order to reduce the chromosome number by one half. This way, when the sperm and egg join during fertilization, the haploid number is reached. The steps of meiosis are as follows:

Major function of Meiosis I - chromosomes are replicated; cells remain diploid.

Prophase I - replicated chromosomes condense and pair with homologues. This forms a tetrad. Crossing over (the exchange of genetic material between homologues to further increase diversity) occurs during Prophase I.
Metaphase I - homologous sets attach to spindle fibers after lining up in the middle of the cell.
Anaphase I - sister chromatids remain joined and move to the poles of the cell.
Telophase I - two new cells are formed, chromosome number is still diploid

Major function of Meiosis II - to reduce the chromosome number in half.

Prophase II - chromosomes condense.

Metaphase II - spindle fibers form again, sister chromatids line up in center of cell, centromeres divide and sister chromatids separate.
Anaphase II - separated chromosomes move to opposite ends of cell.
Telophase II - four haploid cells form for each original sperm germ cell. One viable egg cell gets all the genetic information and three polar bodies form with no DNA. The nuclear membrane reforms and cytokinesis occurs.

Crossing-over, or homologous recombination, is the process where two homologous chromosomes, paired during prophase I, swap alleles of a particular gene. The transfer occurs through the breaking of the chromosome at a particular locus on one chromosome, and the subsequent attachment to the same locus on the paired chromosome. This event is facilitated through cellular machinery.

Homologous recombination is the mechanism by which genetic diversity is conferred. The genotype is altered by the re-assembly of genes on a chromosome; hence the packaging of the genetic material for progeny will differ. As a result, phenotypic changes are observed as the reorganized DNA blueprint imparts different individual traits such as eye color, and is revealed in an organism that exhibits evidence of each parent.

Since it's not a perfect world, mistakes happen. Inheritable changes in DNA are called **mutations**. Mutations may be errors in replication or a spontaneous rearrangement of one or more segments by factors like radioactivity, drugs, or chemicals. The amount of the change is not as critical as where the change is. Mutations may occur on somatic or sex cells. Usually the ones on sex cells are more dangerous since they contain the basis of all information for the developing offspring. Mutations are not always bad. They are the basis of evolution, and if they make a more favorable variation that enhances the organism's survival, then they are beneficial. But, mutations may also lead to abnormalities, birth defects, and even death. There are several types of mutations; let's suppose a normal sequence was as follows:

Normal -   A B C D E F

Duplication - one gene is repeated.   A B C C D E F

Inversion - a segment of the sequence is flipped around.   A E D C B F

Deletion - a gene is left out.   A B C E F

Insertion or Translocation - a segment from another place on the DNA is inserted in the wrong place.   A B C R S D E F

Breakage - a piece is lost.   A B C (DEF is lost)

Nondisjunction – This occurs during meiosis when chromosomes fail to separate properly. One sex cell may get both genes and another may get none. Depending on the chromosomes involved this may or may not be serious. Offspring end up with either an extra chromosome or are missing one. An example of nondisjunction is Down Syndrome, where three of chromosome #21 are present.

Skill 8.4   Demonstrating knowledge of the basic principles of inheritance and Mendel's laws and applying these principles and laws to inheritance problems

Gregor Mendel is recognized as the father of genetics. His work in the late 1800's is the basis of our knowledge of genetics. Although unaware of the presence of DNA or genes, Mendel realized there were factors (now known as genes) that were transferred from parents to their offspring. Mendel worked with pea plants and fertilized the plants himself, keeping track of subsequent generations which led to the Mendelian laws of genetics. Mendel found that two "factors" governed each trait, one from each parent. Traits or characteristics came in several forms, known as alleles. For example, the trait of flower color had white alleles and purple alleles. Mendel formed three laws:

Law of dominance - in a pair of alleles, one trait may cover up the allele of the other trait. Example: brown eyes are dominant to blue eyes.

Law of segregation - only one of the two possible alleles from each parent is passed on to the offspring from each parent. (During meiosis, the haploid number insures that half the sex cells get one allele, half get the other).

Law of independent assortment - alleles sort independently of each other. (Many combinations are possible depending on which sperm ends up with which egg. Compare this to the many combinations of hands possible when dealing a deck of cards).

Definitions:

monohybrid cross - a cross using only one trait.

dihybrid cross - a cross using two traits. More combinations are possible.

Dominant - the stronger of the two traits. If a dominant gene is present, it will be expressed. Shown by a capital letter.

Recessive - the weaker of the two traits. In order for the recessive gene to be expressed, there must be two recessive genes present. Shown by a lower case letter.

Homozygous - (purebred) having two of the same genes present; an organism may be homozygous dominant with two dominant genes or homozygous recessive with two recessive genes.

Heterozygous - (hybrid) having one dominant gene and one recessive gene. The dominant gene will be expressed due to the Law of Dominance.

Genotype - the genes the organism has. Genes are represented with letters. AA, Bb, and tt are examples of genotypes.

Phenotype - how the trait is expressed in an organism. Blue eyes, brown hair, and red flowers are examples of phenotypes.

Incomplete dominance - neither gene masks the other; a new phenotype is formed. For example, red flowers and white flowers may have equal strength. A heterozygote (Rr) would have pink flowers. If a problem occurs with a third phenotype, incomplete dominance is occurring.

Codominance - genes may form new phenotypes. The ABO blood grouping is an example of co-dominance. A and B are of equal strength and O is recessive. Therefore, type A blood may have the genotypes of AA or AO, type B blood may have the genotypes of BB or BO, type AB blood has the genotype A and B, and type O blood has two recessive O genes.

Linkage - genes that are found on the same chromosome usually appear together unless crossing over has occurred in meiosis. (Example - blue eyes and blonde hair)

Lethal alleles - these are usually recessive due to the early death of the offspring. If a 2:1 ratio of alleles is found in offspring, a lethal gene combination is usually the reason. Some examples of lethal alleles include sickle cell anemia, tay-sachs and cystic fibrosis. Usually the coding for an important protein is affected.

Inborn errors of metabolism - these occur when the protein affected is an enzyme. Examples include PKU (phenylketonuria) and albanism.

Polygenic characters - many alleles code for a phenotype. There may be as many as twenty genes that code for skin color. This is why there is such a variety of skin tones. Another example is height. A couple of medium height may have very tall offspring.

Sex linked traits - the Y chromosome found only in males (XY) carries very little genetic information, whereas the X chromosome found in females (XX) carries very important information. Since men have no second X chromosome to cover up a recessive gene, the recessive trait is expressed more often in men. Women need the recessive gene on both X chromosomes to show the trait. Examples of sex linked traits include hemophilia and color-blindness.

Sex influenced traits - traits are influenced by the sex hormones. Male pattern baldness is an example of a sex influenced trait. Testosterone influences the expression of the gene. Mostly men loose their hair due to this trait.

Punnet squares - these are used to show the possible ways that genes combine and indicate probability of the occurrence of a certain genotype or phenotype. One parent's genes are put at the top of the box and the other parent at the side of the box. Genes combine on the square just like numbers that are multiplied in tables we learned in elementary school. P stands for the parental generation, F_1 is the first generation offspring, and F_2 is their offspring.

Example: Monohybrid Cross - four possible gene combinations

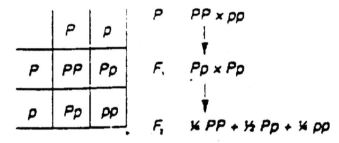

Example: Dihybrid Cross - sixteen possible gene combinations
The genotypes YY and Yy are yellow phenotypically since yellow is dominant; the genotype yy is green which is recessive
Genotypes RR and Rr are dominant round phenotype: rr is wrinkled (recessive)

	YR	Yr	yR	yr
YR	YYRR	YYRr	YyRR	YyRr
Yr	YYRr	YYrr	YyRr	Yyrr
yR	YyRR	YyRr	yyRR	yyRr
yr	YyRr	Yyrr	yyRr	yyrr

P   YYRR x yyrr

↓

F₁   YyRr

↓

F₂   YYRR - 1
      YYRr - 2   } 9 yellow round
      YyRR - 2
      YyRr - 4

      yyRR - 1   } 3 green round
      yyRr - 2

      YYrr - 1   } 3 yellow wrinkled
      Yyrr - 2

      yyrr - 1   } 1 green wrinkled

Skill 8.5   Demonstrating knowledge of applications of the principles of genetics and DNA technology (e.g., selective breeding, forensics, medicine)

In its simplest form, **genetic engineering** requires enzymes to cut DNA, a vector, and a host organism for the recombinant DNA. A **restriction enzyme** is a bacterial enzyme that cuts foreign DNA in specific locations. The restriction fragment that results can be inserted into a bacterial plasmid **(vector)**. Other vectors that may be used include viruses and bacteriophage. The splicing of restriction fragments into a plasmid results in a recombinant plasmid. This recombinant plasmid can now be placed in a host cell, usually a bacterial cell, and replicate.

The use of **recombinant DNA** provides a means to transplant genes among species. This opens the door for cloning specific genes of interest. Hybridization can be used to find a gene of interest. A probe is a molecule complementary in sequence to the gene of interest. The probe, once it has bonded to the gene, can be detected by labeling with a radioactive isotope or a fluorescent tag.

Gel electrophoresis is another method for analyzing DNA. Electrophoresis separates DNA or protein by size or electrical charge. The DNA runs towards the positive charge as it separates the DNA fragments by size. The gel is treated with a DNA-binding dye that fluoresces under ultraviolet light. A picture of the gel can be taken and used for analysis.

One of the most widely used genetic engineering techniques is **polymerase chain reaction (PCR)**. PCR is a technique in which a piece of DNA can be amplified into billions of copies within a few hours. This process requires primer to specify the segment to be copied, and an enzyme (usually tag polymerase) to amplify the DNA. PCR has allowed scientists to perform several procedures on the smallest amount of DNA.

Forensic scientists regularly use DNA technology to solve crimes. DNA testing can determine a person's guilt or innocence. A suspect's DNA "fingerprint" is compared to the DNA found at the crime scene. If the DNA "fingerprints" match, guilt can then be established.

Genetic engineering has made enormous contributions to medicine. Genetic engineering has opened the door to DNA technology. The use of DNA probes and polymerase chain reaction (PCR) has enabled scientists to identify and detect elusive pathogens. Diagnosis of genetic disease is now possible before the onset of symptoms.

Genetic engineering has allowed for the treatment of some genetic disorders. **Gene therapy** is the introduction of a normal allele to the somatic cells to replace the defective allele. The medical field has had success in treating patients with a single enzyme deficiency disease. Gene therapy has allowed doctors and scientists to introduce a normal allele that would provide the missing enzyme.

Insulin and mammalian growth hormones have been produced in bacteria by gene-splicing techniques. Insulin treatment helps control diabetes for millions of people who suffer from the disease. The insulin produced in genetically engineered bacteria is chemically identical to that made in the pancreas. Human grown hormone (HGH) has been genetically engineered for treatment of dwarfism caused by insufficient amounts of HGH. HGH is being further researched for treatment of broken bones and severe burns.

Biotechnology has advanced the techniques used to create vaccines. Genetic engineering allows for the modification of a pathogen in order to attenuate it for vaccine use. In fact, vaccines created by a pathogen attenuated by gene-splicing may be safer than using the traditional mutants.

Many microorganisms are used to detoxify toxic chemicals and to recycle waste. Sewage treatment plants use microbes to degrade organic compounds. Some compounds, like chlorinated hydrocarbons, cannot be easily degraded. Scientists are working on genetically modifying microbes to be able to degrade the harmful compounds that the current microbes cannot.

Genetic engineering has benefited agriculture also. For example, many dairy cows are given bovine growth hormone to increase milk production. Commercially grown plants are often genetically modified for optimal growth.

Strains of wheat, cotton, and soybeans have been developed to resist herbicides used to control weeds. This allows for the successful growth of the plants while destroying the weeds. Crop plants are also being engineered to resist infections and pests. Scientists can genetically modify crops to contain a viral gene that does not affect the plant and will "vaccinate" the plant from a virus attack. Crop plants are now being modified to resist insect attacks. This allows for farmers to reduce the amount of pesticide used on plants.

0009 UNDERSTAND THE DEPENDENCE OF ORGANISMS ON ONE ANOTHER AND UNDERSTAND THE FLOW OF ENERGY AND MATTER IN ECOSYSTEMS

Skill 9.1   Recognizing the characteristics of populations, communities, ecosystems, and biomes

Ecology is the study of organisms, where they live and their interactions with their environment. A **population** is a group of the same species in a specific area. A **community** is a group of populations residing in the same area. Communities that are ecologically similar in regards to temperature, rainfall, and inhabitants are called **biomes**.

Skill 9.2   Analyzing the flow of energy and matter through the abiotic and biotic components of an ecosystem (e.g., carbon cycle, nitrogen cycle)

Essential elements are recycled through an ecosystem. At times, the element needs to be "fixed" in a useable form. Cycles are dependent on plants, algae and bacteria to fix nutrients for use by other animals.

> **Water cycle** - 2% of all the available water is fixed and held in ice or the bodies of organisms. Available water includes surface water (lakes, ocean, and rivers) and ground water (aquifers, wells). 96% of all available water is from ground water. Water is recycled through the processes of evaporation and precipitation. The water present now is the water that has been here since our atmosphere formed.

> **Carbon cycle** - Ten percent of all available carbon in the air (from carbon dioxide gas) is fixed by photosynthesis. Plants fix carbon in the form of glucose, animals eat the plants and are able to obtain their source of carbon. When animals release carbon dioxide through respiration, the plants again have a source of carbon to fix.

> **Nitrogen cycle** - Eighty percent of the atmosphere is in the form of nitrogen gas. Nitrogen must be fixed and taken out of the gaseous form to be incorporated into an organism. Only a few genera of bacteria have the correct enzymes to break the triple bond between nitrogen atoms. These bacteria live within the roots of legumes (peas, beans, alfalfa) and add bacteria to the soil so it may be taken up by the plant. Nitrogen is necessary to make amino acids and the nitrogenous bases of DNA.

Phosphorus cycle - Phosphorus exists as a mineral and is not found in the atmosphere. Fungi and plant roots have structures called mycorrhizae that are able to fix insoluble phosphates into useable phosphorus. Urine and decayed matter returns phosphorus to the earth where it can be fixed in the plant. Phosphorus is needed for the backbone of DNA and for the manufacture of ATP.

Skill 9.3   Analyzing the symbiotic relationships among organisms in an ecosystem

The term ecosystem is used to refer to the biotic and abiotic components and processes that comprise a particular subset of the biosphere. An ecosystem encompasses the resources, habitats, flora and fauna of an area. Living organisms of an ecosystem are all interdependent.

Symbiosis is a close association and co-existence of two or more organisms of an ecosystem or community over a prolonged period of time. Symbiosis usually occurs between organisms that are phylogenetically unrelated and the relationship benefits at least one of the organisms. The general concept of symbiosis can be divided into three categories: parasitism, mutualism, and commensalism.

Parasitism is a one-sided relationship of dependence in which one of the symbionts (the parasite) benefits at the expense of the other symbiont (the host). The parasite is usually the smaller of the two organisms. The heartworm is a parasitic roundworm that can be found in wild animals such as wolves, coyotes and foxes, as well as in household pets. The heartworm is spread to its host through mosquito bites, and derives its name during the final stages of its life cycle, when the heartworm resides in the heart of its host and can induce congestive heart failure.

Mutualism is a symbiotic relationship in which both symbionts derive reciprocal benefit from the association. A classic example of mutualism occurs between the clownfish and the sea anemone. Clownfish are a species of brightly colored reef fish found in the Pacific and Indian Oceans. Sea anemones consist of hollow cylinders surrounded by tentacles equipped with nematocysts, specialized cells that contain poison to paralyze or kill prey. The clownfish secretes a mucous that causes the sea anemone to treat the clownfish as if it were part of the same organism. Consequently, the clownfish is not stung by the anemone. By living among the anemone's tentacles, the clownfish is protected from predation by predators not immune to the anemone's sting. The clownfish's bright colors attract other fish to the sea anemone, which then stings and consumes the other fish, allowing the clownfish to feed on the remains. Both species, therefore, benefit from this interaction.

Another example of mutualism is *cleaning symbiosis*, commonly occurring between small cleaner fish, usually marked with horizontal stripes, and large fish of the same or differing species. Large fish approach reefs where cleaner fish reside, and remain stationary while cleaner fish remove parasites, dead skin, old tissue and mucous. Cleaner fish even enter into the mouths and gills of the larger fish without fear of predation. The cleaning of the larger fish serves to maintain its health, and in return, the small cleaner fish are able to feed and receive protection.

Commensalism is the term used to describe the symbiotic association in which organisms share resources without harming one another. For true commensalism to occur, the first species must benefit from the relationship without affecting the second. Commensalism can be seen in the relationship of the pseudoscorpion, a small, predaceous arthropod, and several species of large beetle. Pseudoscorpions are usually less than one centimeter in length, and disperse by concealing themselves under the wing covers of large beetles such as the cerambycid beetle. The pseudoscorpions gain the advantage of being dispersed over wide areas while being protected from predation by the large beetle's size. The beetle, however, is unaffected by the presence of the scorpion.

Skill 9.4   Analyzing the flow of energy through food chains and food webs

Competition - two species that occupy the same habitat or eat the same food are said to be in competition with each other.

Predation - animals that eat other animals are called predators. The animals they feed on are called the prey. Population growth depends upon competition for food, water, shelter, and space. The amount of predators determines the amount of prey, which in turn affects the number of predators. For example, if the coyote population increases, they eat more rabbits which decreases the rabbit population. If there are fewer rabbits, then some of the coyotes will starve and their population will decrease. When the coyotes decrease, then more rabbits will survive.

Trophic levels are based on the feeding relationships that determine energy flow and chemical cycling.

Autotrophs are the primary producers of the ecosystem. **Producers** mainly consist of plants. **Primary consumers** are the next trophic level. The primary consumers are the herbivores that eat plants or algae. **Secondary consumers** are the carnivores that eat the primary consumers. **Tertiary consumers** eat the secondary consumer. These trophic levels may go higher depending on the ecosystem. **Decomposers** are consumers that feed off animal waste and dead organisms. This pathway of food transfer is known as the food chain.

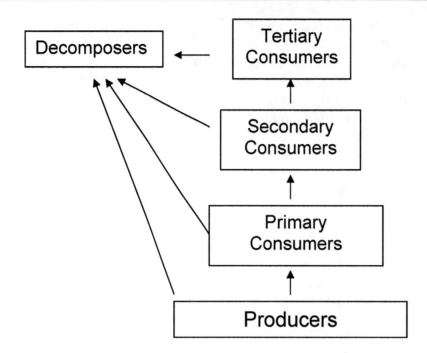

Most food chains are more elaborate, becoming food webs.

Skill 9.5   Analyzing factors that affect population dynamics in an ecosystem (e.g., carrying capacity, human activities)

Biotic factors - living things in an ecosystem; plants, animals, bacteria, fungi, etc. If one population in a community increases, it affects the ability of another population to succeed by limiting the available amount of food, water, shelter and space.

Abiotic factors - non-living aspects of an ecosystem; soil quality, rainfall, and temperature. Changes in climate and soil can cause effects at the beginning of the food chain, thus limiting or accelerating the growth of populations.

Carrying Capacity - this is the total amount of life a habitat can support.   Once the habitat runs out of food, water, shelter, or space, the carrying   capacity decreases, and then stabilizes.

Ecological Problems - nonrenewable resources are fragile and must be conserved for use in the future. Man's impact and knowledge of conservation will control our future.

Biological magnification - chemicals and pesticides accumulate along the food chain. Tertiary consumers have more accumulated toxins than animals at the bottom of the food chain.

Simplification of the food web - Three major crops feed the world (rice, corn, wheat). The planting of these foods wipe out habitats and push animals residing there into other habitats causing overpopulation or extinction.

Fuel sources - strip mining and the overuse of oil reserves have depleted these resources. At the current rate of consumption, conservation or alternate fuel sources will guarantee our future fuel sources.

Pollution - although technology gives us many advances, pollution is a side effect of production. Waste disposal and the burning of fossil fuels have polluted our land, water and air. Global warming and acid rain are two results of the burning of hydrocarbons and sulfur.

Global warming - rainforest depletion and the use of fossil fuels and aerosols have caused an increase in carbon dioxide production. This leads to a decrease in the amount of oxygen, which is directly proportional to the amount of ozone. As the ozone layer depletes, more heat enters our atmosphere and is trapped. This causes an overall warming effect, which may eventually melt polar ice caps, causing a rise in water levels and changes in climate, which will affect weather systems world-wide.

Endangered species - construction of homes to house people in our overpopulated world has caused the destruction of habitat for other animals leading to their extinction.

Overpopulation - the human race is still growing at an exponential rate. Carrying capacity has not been met due to our ability to use technology to produce more food and housing. Space and water can not be manufactured and eventually our non-renewable resources will reach a crisis state. Our overuse affects every living thing on this planet.

Skill 9.6   Identifying characteristics of the earth's major terrestrial biomes and aquatic communities

Specific biomes include:

Marine - covers 75% of the earth. This biome is organized by the depth of the water. The intertidal zone is from the tide line to the edge of the water. The littoral zone is from the water's edge to the open sea. It includes coral reef habitats and is the most densely populated area of the marine biome. The open sea zone is divided into the epipelagic zone and the pelagic zone. The epipelagic zone receives more sunlight and has a larger number of species. The ocean floor is called the benthic zone and is populated with bottom feeders.

Tropical Rain Forest - temperature is constant (25 degrees C), rainfall exceeds 200 cm. per year. Located around the area of the equator, the rain forest has abundant, diverse species of plants and animals.

Savanna - temperatures range from 0-25 degrees C depending on the location. Rainfall is from 90 to 150 cm per year. Plants include shrubs and grasses. The savanna is a transitional biome between the rain forest and the desert.

Desert - temperatures range from 10-38 degrees C. Rainfall is under 25 cm per year. Plant species include xerophytes and succulents. Lizards, snakes and small mammals are common animals.

Temperate Deciduous Forest - temperature ranges from -24 to 38 degrees C. Rainfall is between 65 to 150 cm per year. Deciduous trees are common, as well as deer, bear and squirrels.

Taiga - temperatures range from -24 to 22 degrees C. Rainfall is between 35 to 40 cm per year. Taiga is located very north and very south of the equator, getting close to the poles. Plant life includes conifers and plants that can withstand harsh winters. Animals include weasels, mink, and moose.

Tundra - temperatures range from -28 to 15 degrees C. Rainfall is limited, ranging from 10 to 15 cm per year. The tundra is located even further north and south than the taiga. Common plants include lichens and mosses. Animals include polar bears and musk ox.

Polar or Permafrost - temperature ranges from -40 to 0 degrees C. It rarely gets above freezing. Rainfall is below 10 cm per year. Most water is bound up as ice. Life is limited.

0010 UNDERSTAND THE THEORY OF EVOLUTION AND THE ROLE OF NATURAL SELECTION

Skill 10.1 **Demonstrating knowledge of the historical development of the theory of evolution and supporting evidence (e.g., Darwin's finches)**

The idea of evolution is actually an ancient one, but scientists typically credit the the first plausible theory of evolution to Charles Darwin and Alfred Russel Wallace. These scientists jointly presented the theory of evolution by natural selection in 1858. Earlier Jean Baptiste Lamarck had watched giraffes and felt that they acquired their long necks through stretching. He then hypothesized that they passed on their acquired longer necks to their descendants. Although his mechanism was incorrect, he had introduced the idea of species changing over time by some natural process.

One of the more famous historical incidents leading up the formulation of Darwin's theory was his travel aboard the HMS Beagle from 1831 to 1836. During the Beagle's 5-year journey across the Atlantic and along the coast of South America, Darwin made many observations regarding the plant and animal life. He later collected and published these findings, which were the basis for the theory he later developed, in "The Origin of the Species." His findings were presented at the same meeting of the Linnaean Society in London as those of Alfred Russel Wallace. Wallace was a naturalist who had been collecting plants and animals in Malaysia and Indonesia and had conceived the idea of evolution through differential reproduction.

Since the development of the theory of evolution, some of the best supporting evidence has come from molecular genetic studies. However, during the time of Darwin there were several "lower tech" pieces of evidence. For instance, the fossil record and comparative anatomy demonstrated possible relationships between both living and extinct organisms. One of the most famous examples of this involves "Darwin's finches." While onboard the Beagle, Darwin observed 13 species of finches in the Galapagos Islands. These finches were similar in size, but had dramatically different sized and shaped beaks and extremely varied behavior. While Darwin did not at first realize that all these birds were related species of finches, their relations became apparent when he returned specimens to England and consulted with other naturalists. He later discovered that these were all related species that had adapted to take advantage of different food sources and slightly different ecological niches.

Although Darwin is credited with "the theory of evolution," there are 20+ other scientists who have purported various theories or variations on theories of evolution. Many scientists accept the modern synthetic theory of evolution which combines Darwinism with modern genetics and molecular biology, but they may have different opinions of mechanisms of speciation. Such controversy is all part of the scientific process.

Skill 10.2   Recognizing the relationship between natural selection and adaptation

Darwin defined the Theory of Natural Selection in the mid-1800's. Through the study of finches on the Galapagos Islands, Darwin theorized that nature selects the traits that are advantageous to the organism. Those that do not possess the desirable trait die and do not pass on their genes. Those more fit to survive reproduce, thus increasing that gene in the population. Darwin listed four principles to define natural selection:

1. The individuals in a certain species vary from generation to generation.
2. Some of the variations are determined by the genetic makeup of the species.
3. More individuals are produced than will survive to maturity.
4. Some genes allow for better survival of an animal.

Causes of evolution - Certain factors increase the chances of variability in a population, thus leading to evolution. Items that increase variability include mutations, sexual reproduction, immigration, and large population. Items that decrease variation would be natural selection, emigration, small population, and random mating.

Anatomical structures and physiological processes that evolve over geological time to increase the overall reproductive success of an organism in its environment are known as biological **adaptations**. Such evolutionary changes occur through natural selection, the process by which individual organisms with favorable traits survive to reproduce more frequently than those with unfavorable traits. The heritable component of such favorable traits is passed down to offspring during reproduction, increasing the frequency of the favorable trait in a population over many generations.

Adaptations increase long-term reproductive success by making an organism better suited for survival under particular environmental conditions and pressures. These biological changes can increase an organism's ability to obtain air, water, food and nutrients and to cope with environmental variables as well as to defend themselves. The term adaptation may apply to changes in biological processes that, for example, enable on organism to produce venom or to regulate body temperature, and also to structural adaptations, such as an organisms' skin color and shape. Adaptations can occur in behavioral traits and survival mechanisms as well.

One well-known structural change that demonstrates the concept of adaptation is the development of the primate and human opposable thumb, the first digit of the hand that can be moved around to touch other digits and to grasp objects. The history of the opposable thumb is one of complexly linked structural and behavioral adaptations in response to environmental stressors.

Skill 10.3   Analyzing the roles of variation, natural selection, nonrandom selection, and isolation in evolutionary change and speciation

The most commonly used species concept is the **Biological Species Concept (BSC)**. This states that a species is a reproductive community of populations that occupy a specific niche in nature. It focuses on <u>reproductive isolation</u> of populations as the primary criterion for recognition of species status. The biological species concept does not apply to organisms that are completely asexual in their reproduction, fossil organisms, or distinctive populations that hybridize.

Reproductive isolation is caused by any factor that impedes two species from producing viable, fertile hybrids. Reproductive barriers can be categorized as **prezygotic** (premating) or **postzygotic** (postmating).

The prezygotic barriers are as follows:

1. Habitat isolation – species occupy different habitats in the same territory.
2. Temporal isolation – populations reaching sexual maturity or flowering at different times of the year.
3. Ethological isolation – behavioral differences that reduce or prevent interbreeding between individuals of different species (including pheromones and other attractants).
4. Mechanical isolation – structural differences that make gamete transfer difficult or impossible.
5. Gametic isolation – male and female gametes do not attract each other; no fertilization.

The postzygotic barriers are as follows:

1. Hybrid inviability – hybrids die before sexual maturity.
2. Hybrid sterility – disrupts gamete formation; no normal sex cells.
3. Hybrid breakdown – reduces viability or fertility in progeny of the F_2 backcross.

Geographical isolation can also lead to the origin of species. **Allopatric speciation** is speciation without geographic overlap. It is the accumulation of genetic differences through division of a species' range, either through a physical barrier separating the population or through expansion by dispersal such that gene flow is cut. In **sympatric speciation**, new species arise within the range of parent populations. Populations are sympatric if their geographical range overlaps. This usually involves the rapid accumulation of genetic differences (usually chromosomal rearrangements) that prevent interbreeding with adjacent populations.

Sexual selection - Genes that happen to come together determine the makeup of the gene pool. Animals that use mating behaviors may be successful or unsuccessful. An animal that lacks attractive plumage or has a weak mating call will not attract the female, thereby eventually limiting that gene in the gene pool. Mechanical isolation, where sex organs do not fit the female, has an obvious disadvantage.

Skill 10.4   Identifying evidence for evolutionary change in organisms over time and for evolutionary relationships among organisms (e.g., fossils, biochemical similarities*)*

Fossils are the key to understanding biological history. They are the preserved remnants left by an organism that lived in the past.   Scientists have established a geological time scale to determine the age of a fossil. The geological time scale is broken down into four eras: the Precambrian, Paleozoic, Mesozoic, and Cenozoic. The eras are further broken down into periods that represent a distinct age in the history of Earth and its life. Scientists use rock layers called strata to date fossils. The older layers of rock are at the bottom. This allows scientists to correlate the rock layers with the era they date back to. Radiometric dating is a more precise method of dating fossils. Rocks and fossils contain isotopes of elements accumulated over time. The isotope's half-life is used to date older fossils by determining the amount of isotope remaining and comparing it to the half-life.

Dating fossils is helpful to construct an evolutionary tree. Scientists can arrange the succession of animals based on their fossil record. The fossils of an animal's ancestors can be dated and placed on its evolutionary tree. For example, the branched evolution of horses shows the progression of the modern horse's ancestors to be larger, to have a reduced number of toes, and have teeth modified for grazing.

Molecular genetics is the study of the structure and function of genes at the molecular level. The genetic structures and DNA sequences of an organism reveal the organism's evolutionary history. Scientists use tools of molecular genetics to study mutations in DNA that can lead to natural selection and evolution. The study of DNA using molecular genetic techniques provides us with statistics such as the 95% similarity between humans and chimpanzees.

There are many observations and examples in molecular genetics that show evolutionary relationships between organisms. For example, the amino acid sequence of Cytochrome c, a respiratory pigment found in eukaryotic cells, has changed slowly over time. Thus, when comparing two organisms, the amount of difference in the amino acid sequence of Cytochrome c estimates the degree of evolutionary relationship. The smaller the difference the closer the relationship between the organisms. Humans and chimpanzees have identical Cytochrome c sequences, the difference between humans and rhesus monkeys is a single amino acid, the difference between humans and penguins is 11 amino acids, and the difference between humans and yeast is 38 amino acids. Such comparisons support the evolutionary theory that small changes in DNA sequence can lead to large diversions in lineages.

Another observation in molecular genetics related to evolutionary theory is the origin of mitochondria in eukaryotic cells. Studies and comparisons of mitochondrial DNA and bacterial DNA reveal a close relationship. From these observations, many scientists believe that mitochondria originated from free-living bacteria.

0011   UNDERSTAND THE NATURE OF MATTER AND ITS CLASSIFICATION

Skill 11.1   Demonstrating knowledge of the structure, properties, and forces within the atom and the historical development of theories of atomic structure

The **atomic theory of matter** suggests that:

1. All matter consists of atoms
2. All atoms of an element are identical
3. Different elements have different atoms
4. Atoms maintain their properties in a chemical reaction

The atomic theory of matter was first suggested by a Greek named **Democritus.** The atomic theory of matter states that matter is made up of tiny, rapidly moving particles. These particles move more quickly when warmer, because temperature is a measure of average kinetic energy of the particles. Warmer molecules therefore move further away from each other, with enough energy to separate from each other more often and for greater distances.

Much later (1803-1807), a scientist named **John Dalton** expanded on Democritus' idea. Dalton, a school teacher, made some observations about air: air is a mixture of different kinds of gases; these gases do not separate on their own; it is possible to compress gases into a smaller volume. He also thought that particles of different substances must be different from each other and must maintain their own mass when combined with other substances.

Dalton's Model of the Atom:

1. Matter is made up of small particles called atoms.
2. Atoms of an element are identical to each other in mass and other properties..
3. Atoms of different elements have different masses and differ from each other in other characteristics.
4. Atoms of different elements combine with each other to form compounds.
5. Atoms of an element are not created, destroyed or changed into a different type of atom by chemical reactions.
6. In a given compound, the relative numbers and kinds of atoms are constant.

The present model of the atom is much different from Dalton's model. His model for the atom was a tiny, indivisible, indestructible particle.

In the late 1800's, a British scientist named **J. J. Thomson** was studying how electric current flowed through a vacuum tube. His hypothesis was:

1. If rays are made of charged particles, then an electric field would attract them.
2. If it is a charged particle, then a magnet will affect its motion.

From his work, Thomson proved that the rays were made of negative particles. These particles were later called electrons.

The results of his experimentation produced **Thomson's Model:** The atom is made of negative particles equally mixed in a sphere of positive material. His model has been referred to as the "plum-pudding" model of the atom with the electrons as plums in a pudding of positive electrons.

In 1886 it was discovered that some elements give off particles with a positive charge. These elements have 7,000 times the mass of electrons. The British scientist **Ernest Rutherford** called these **alpha particles**. He used the alpha particles to test Thompson's model. He hammered gold foil until it was less than 1mm thick and then fired alpha particles at the foil. He used a telescope and a screen to locate the alpha particles. His hypothesis was that if Thomson's theory was right, then the alpha particles would pass through the foil in a straight line. He found that most particles passed through as expected. However, some appeared to bounce off in another direction. This could not be explained by Thomson's model. The result of his experiment gave way to **Rutherford's Model**:

1. Most of the atom is empty space. (This explains why most of the alpha particles pass directly through it).
2. The center of the atom contains a nucleus containing most of the mass and all of the positive charge of the atom.
3. The scattering of particles occurs when they collide with the nucleus.
4. The region of the space outside the nucleus is occupied by electrons.
5. The atom is neutral because the protons in the nucleus equal the electrons in the space outside the nucleus.

Based on Rutherford's model, scientists thought that the electrons of an atom might orbit the nucleus much like the planets orbit the sun. If this is true, they could expect two things:

1. As electrons orbit, they give off light energy continuously. If this light energy is passed through a prism, it would produce a band of color.
2. As the orbiting electrons gave off light, they would lose energy and spiral into the nucleus of the atom causing the atom to collapse. Therefore, the atom would take up no space.

No color band was observed. Instead, lines of color and dark lines were observed. Also, since we know that because matter does in fact take up space, then the orbiting atoms can not collapse into nothing. Another model was necessary to explain the observations. The Danish scientist **Neils Bohr** created a model in 1913. The results of his model are:

1. Electrons orbit the nucleus, but only certain orbits are allowed. An electron in an allowed orbit will not lose energy.
2. When an electron moves from an outer orbit to an inner orbit, it gives off energy.
3. When an electron moves from an inner orbit to an outer orbit, it absorbs energy.

Bohr's model only explains the very simplest atoms, such as hydrogen. It was further refined. Louis De Broglie's model of the atom described electrons as matter waves in standing wave orbits around the nucleus. Werner Heisenberg's Uncertainty Principle applies to the location and momentum of electrons in an atom. Erwin Schrodinger described probable orbitals for electrons and electron densities. **Wolfgang Pauli** helped develop quantum mechanics in the 1920s by developing the concept of spin and the **Pauli exclusion principle**, which states that if two electrons occupy the same orbital, they must have different spin (intrinsic angular momentum). This principle has been generalized to other quantum particles. **Friedrich Hund** determined a set of **rules to determine the ground state** of a multi-electron atom in the 1920s. One of these rules is called **Hund's Rule** in introductory chemistry courses, and describes the order in which electrons fill orbitals and their spin.

An **atom** is a nucleus surrounded by a cloud with moving electrons.

The **nucleus** is the center of the atom. The positive particles inside the nucleus are called **protons.** The mass of a proton is about 2,000 times that of the mass of an electron. The number of protons in the nucleus of an atom is called the **atomic number**. All atoms of the same element have the same atomic number.

Neutrons are another type of particle in the nucleus. Neutrons and protons have about the same mass, but neutrons have no charge. Neutrons were discovered because scientists observed that not all atoms in neon gas have the same mass. They had identified isotopes. **Isotopes** of an element have the same number of protons in the nucleus, but have different masses. Neutrons explain the difference in mass. They have mass but no charge.

The mass of matter is measured against a standard mass such as the gram. Scientists measure the mass of an atom by comparing it to that of a standard atom. The result is relative mass. The **relative mass** of an atom is its mass expressed in terms of the mass of the standard atom. The isotope of the element carbon is the standard atom. It has six (6) neutrons and is called carbon-12. It is assigned a mass of 12 atomic mass units (amu). Therefore, the **atomic mass unit (amu)** is the standard unit for measuring the mass of an atom. It is equal to the mass of a carbon atom.

The **mass number** of an atom is the sum of its protons and neutrons. In any element, there is a mixture of **isotopes**, some having slightly more or slightly fewer neutrons. The **average atomic mass** of an element is an average of the mass numbers of its isotopes.

The following table summarizes the terms used to describe atomic nuclei:

Term	Example	Meaning	Characteristic
Atomic Number	# protons (p)	same for all atoms of a given element	Carbon (C) atomic number = 6 (6p)
Mass number	# protons + # neutrons (p + n)	changes for different isotopes of an element	C-12 (6p + 6n) C-13 (6p + 7n)
Atomic mass	average mass of the atoms of the element	usually not a whole number	atomic mass of carbon equals 12.011

Each atom has an equal number of electrons (negative) and protons (positive). Therefore, atoms are neutral electrically. Electrons orbiting the nucleus occupy energy levels that are arranged in order and the electrons tend to occupy the lowest energy level available. A **stable electron arrangement** is an atom that has all of its electrons in the lowest possible energy levels.

Each energy level holds a maximum number of electrons. However, an atom with more than one level does not hold more than 8 electrons in its outermost energy level.

Level	Name	Max. # of Electrons
First	K	2
Second	L	8
Third	M	18
Fourth	N	32

This can help explain why chemical reactions occur. Atoms react with each other when their outer levels are unfilled. When atoms either exchange or share electrons with each other, these energy levels become filled and the atom becomes more stable.

As an electron gains energy, it moves from one energy level to a higher energy level. The electron cannot leave one level until it has enough energy to reach the next level. **Excited electrons** are electrons that have absorbed energy and have moved farther from the nucleus.

Electrons can also lose energy. When they do, they fall to a lower level. However, they can only fall to the lowest level that has room for them. This explains why atoms do not collapse.

Skill 11.2   Distinguishing among types of matter (e.g., elements, compounds, mixtures)

An **element** is a substance that can not be broken down into other substances. To date, scientists have identified 109 elements: 89 are found in nature and 20 are synthetic.

An **atom** is the smallest particle of the element that retains the properties of that element. All of the atoms of a particular element are the same. The atoms of each element are different from the atoms of other elements.

Elements are assigned an identifying symbol of one or two letters. The symbol for oxygen is O and stands for one atom of oxygen. However, because oxygen atoms in nature are joined together is pairs, the symbol O_2 represents oxygen. This pair of oxygen atoms is a molecule. A **molecule** is the smallest particle of substance that can exist independently and has all of the properties of that substance. A molecule of most elements is made up of one atom. However, oxygen, hydrogen, nitrogen, bromine, iodine, fluorine, and chlorine molecules are made of two atoms each so they are known as diatomic.

A **compound** is made of two or more elements that have been chemically combined. Atoms join together when elements are chemically combined. The result is that the elements lose their individual identities when they are joined. The compound that they become has different properties.

A compound is a pure substance formed from the combination of two or more elements that differs from the elements in it. Compounds obey the **law of definite proportions** which says that the elements in a compound always combine in the same proportion by mass. All compounds are electrically neutral. There are two types of compounds: **ionic compounds** are formed using ions and use electrostatic charge and attraction while **molecular compounds** are formed of covalently bonded atoms. The smallest unit of either one is sometimes referred to as a **molecule**; however, the molecule is the smallest unit of a molecular compound while a **formula unit** is the smallest unit of an ionic compound.

We use a formula to show the elements of a chemical compound. A **chemical formula** is a shorthand way of showing what is in a compound by using symbols and subscripts. The letter symbols let us know what elements are involved and the number subscript tells how many atoms of each element are involved. No subscript is used if there is only one atom involved. For example, carbon dioxide is made up of one atom of carbon (C) and two atoms of oxygen (O_2), so the formula would be represented as CO_2.

Substances can combine without a chemical change. A **mixture** is any combination of two or more elements or substances in which the substances keep their own properties. A fruit salad is a mixture. So is an ice cream sundae, although you might not recognize each part if it is stirred together. Colognes and perfumes are the other examples. You may not readily recognize the individual elements. However, they can be separated.

Compounds and **mixtures** are similar in that they are made up of two or more substances. However, they have the following opposite characteristics:

Compounds:
1. Made up of one kind of particle
2. Formed during a chemical change
3. Broken down only by chemical changes
4. Properties are different from its parts
5. Has a specific amount of each ingredient.

Mixtures:
1. Made up of two or more particles
2. Not formed by a chemical change
3. Can be separated by physical changes
4. Properties are the same as its parts.
5. Does not have a definite amount of each ingredient.

Skill 11.3   Analyzing physical and chemical properties of matter (e.g., malleability, melting point, reactivity)

Everything in our world is made up of **matter**, whether it is a rock, a building, an animal, or a person. Matter is defined by its characteristics. All matter has physical properties and chemical properties. Physical properties and chemical properties of matter describe the appearance or behavior of a substance. A **physical property** can be observed without changing the identity of a substance. For instance, you can describe the color, mass, shape, and volume of a book. Physical properties include color, size, luster, smell, freezing point, boiling point, melting point, mass, and density. **Chemical properties** describe the ability of a substance to be changed into new substances. Chemical properties include such things as heat of combustion, reactivity with water, pH, and electromotive force. Baking powder goes through a chemical change as it changes into carbon dioxide gas during the baking process. Measuring physical properties does not change the matter, but measuring the chemical properties involves a chemical change to the matter.

Physical Properties:

Matter takes up space and it has mass. **Mass** is a measure of the amount of matter in an object. Two objects of equal mass will balance each other on a simple balance scale no matter where the scale is located. For instance, two rocks with the same amount of mass that are in balance on earth will also be in balance on the moon. They will feel heavier on earth than on the moon because of the gravitational pull of the earth. So, although the two rocks have the same mass, they will have different **weight.**

Weight is the measure of the earth's pull of gravity on an object. It can also be defined as the pull of gravity between other bodies. The units of weight measurement commonly used are the pound (English measure) and the kilogram (metric measure).

In addition to mass, matter also has the property of volume. **Volume** is the amount of cubic space that an object occupies. Volume and mass together give a more exact description of the object. Two objects may have the same volume, but different mass, or the same mass but different volumes, etc. For instance, consider two cubes that are each one cubic centimeter, one made from plastic, one from lead. They have the same volume, but the lead cube has more mass. The measure that we use to describe the cubes takes into consideration both the mass and the volume. **Density** is the mass of a substance contained per unit of volume. If the density of an object is less than the density of a liquid, the object will float in the liquid. If the object is denser than the liquid, then the object will sink.

Density is stated in grams per cubic centimeter (g/cm^3) where the gram is the standard unit of mass. To find an object's density, you must measure its mass and its volume. Then divide the mass by the volume (D = m/V).

To discover an object's density, first use a balance to find its mass. Then calculate its volume. If the object is a regular shape, you can find the volume by multiplying the length, width, and height together. However, if it is an irregular shape, you can find the volume by seeing how much water it displaces. Measure the water in the container before and after the object is submerged. The difference will be the volume of the object.

Specific gravity is the ratio of the density of a substance to the density of water. For instance, the specific density of one liter of turpentine is calculated by comparing its mass (0.81 kg) to the mass of one liter of water (1 kg):

$$\frac{\text{mass of 1 L alcohol}}{\text{mass of 1 L water}} = \frac{0.81\ \text{kg}}{1.00\ \text{kg}} = 0.81$$

Conductivity: Substances can have two variables of conductivity. A conductor is a material that transfers a substance easily. That substance may be thermal or electrical in nature. Metals are known for being good thermal and electrical conductors. Touch your hand to a hot piece of metal and you know it is a good conductor- the heat transfers to your hand and you may be burned. Materials through which electric charges can easily flow are called electrical **conductors**. Metals that are good electric conductors include silicon and boron. On the other hand, an **insulator** is a material through which electric charges do not move easily, if at all. Examples of electrical insulators would be the nonmetal elements of the periodic table.

Solubility is defined as the amount of substance (referred to as solute) that will dissolve into another substance, called the solvent. The amount that will dissolve can vary according to the conditions, most notably temperature. The process is called solvation.

Melting point refers to the temperature at which a solid becomes a liquid. Melting takes place when there is sufficient energy available to break the intermolecular forces that hold molecules together in a solid.

Boiling point refers to the temperature at which a liquid becomes a gas. Boiling occurs when there is enough energy available to break the intermolecular forces holding molecules together as a liquid.

Hardness describes how difficult it is to scratch or indent a substance. The hardest natural substance is diamond.

Chemical Properties:

Heat of combustion is the heat of reaction that is evolved by the complete combustion of one mole of a substance.

Because hydrogen has the potential to ignite and explode, it has **explosiveness** or **flammability** as a chemical property.

Some elements such as sodium are highly **reactive with water**.

The **pH** is a measure of the **acidity** or **basicity** of a solution.

Skill 11.4   Demonstrating knowledge of the different types of chemical bonds (i.e., ionic, covalent, and metallic) and how the character of chemical bonds affects the properties of substances

The outermost electrons in the atoms are called **valence electrons.** Because they are the ones involved in the bonding process, they determine the properties of the element.

A **chemical bond** is a force of attraction that holds atoms together. When atoms are bonded chemically, they cease to have their individual properties. For instance, hydrogen and oxygen combine into water and no longer look like hydrogen and oxygen. They look like water.

A **covalent bond** is formed when two atoms share electrons. Recall that atoms whose outer levels are not filled with electrons are unstable. When they are unstable, they readily combine with other unstable atoms. By combining and sharing electrons, they act as a single unit. Covalent bonding happens among nonmetals. Covalent bonds are always nonpolar between two non-identical atoms. If the bonding electrons are shared unequally, the bond is polar and the amount of polarity is determined by the difference of electronegativities between the two elements.

Covalent compounds are compounds whose atoms are joined by covalent bonds. Table sugar, methane, and ammonia are examples of covalent compounds.

An **ionic bond** is a bond formed by the transfer of electrons. It happens when metals and nonmetals bond. Before chlorine and sodium combine, the sodium has one valence electron and chlorine has seven. Neither outermost valence level is filled, but the chlorine's outermost level is almost full. During the reaction, the sodium gives one valence electron to the chlorine atom. Both atoms then have filled shells and are stable. Something else happens during the bonding. Before the bonding, both atoms were neutral. When one electron was transferred, it upset the balance of protons and electrons in each atom. The chlorine atom took on one extra electron and the sodium atom released one atom. The atoms have now become ions. **Ions** are atoms with an unequal number of protons and electrons. To determine whether the ion is positive or negative, compare the number of protons (+charge) to the electrons (-charge). If there are more electrons the ion will be negative. If there are more protons, the ion will be positive.

Compounds that result from the transfer of metal atoms to nonmetal atoms are called **ionic compounds.** Sodium chloride (table salt), sodium hydroxide (drain cleaner), and potassium chloride (salt substitute) are examples of ionic compounds.

Metallic bonds are present when the bonding is between two metals. Metallic properties, such as low ionization energies, conductivity, and malleability, suggest that metals possess strong forces of attraction between atoms, but still have electrons that are able to move freely in all directions throughout the metal. This creates a "sea of electrons" model where electrons are quickly and easily transferred between metal atoms. In this model, the outer shell electrons are free to move. The metallic bond is the force of attraction that results from the moving electrons and the positive nuclei left behind. The strength of metal bonds usually results in regular structures and high melting and boiling points.

Skill 11.5 Applying knowledge of the organization of the Periodic Table and its trends to support understanding of/aid in determining the structure and properties of matter

The **periodic table of elements** is an arrangement of the elements in rows and columns so that it is easy to locate elements with similar properties. The elements of the modern periodic table are arranged in numerical order by atomic number. The Periodic Table of the Elements used today is based on the one created by Dmitri Mendeleev who saw that many of the elements were metals and that most of the elements could be grouped according to various properties.

Each box of the Periodic Table shows at least four pieces of data: (1) the element symbol represented by one or two letters, the first letter is capitalized with the second one being lower case, (2) the element name, (3) the atomic number, and (4) the atomic weight.

The **periods** are the rows across the table. They are called first period, second period, etc. The columns of the periodic table are called **groups**, or **families.** Elements in a family have similar properties.

There are three types of elements that are grouped by color: metals, nonmetals, and metalloids.

Element Key

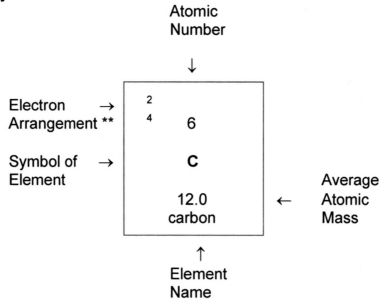

** Number of electrons on each level. Top number represents the innermost level.

The periodic table arranges metals into families with similar properties. The periodic table has its columns marked IA - VIIIA. These are the traditional group numbers. Arabic numbers 1 - 18 are also used, as suggested by the Union of Physicists and Chemists. The Arabic numerals will be used in this text.

Metals:

Metals normally have one to three electrons in the outermost level. Therefore, they easily give up those electrons to achieve a stable outmost electron level, and thus become cations (positively charged ions).

With the exception of hydrogen, all elements in Group 1 are **alkali metals**. These metals are shiny, hard, conduct heat and electricity, malleable, ductile, and are the most chemically active. They have one electron in their outermost level.

Group 2 metals are the **alkaline earth metals.** They are harder, denser, have higher melting points, are usually malleable and ductile, are conductors, and are chemically active. They have two electrons in their outmost level.

The **transition elements** can be found by finding the periods (rows) from 4 to 7 under the groups (columns) 3 - 12. They are metals that do not show a range of properties as you move across the chart. They are hard and have high melting points. Compounds of these elements are colorful, such as silver, gold, and mercury.

Elements can be combined to make metallic objects. An **alloy** is a mixture of two or more elements having properties of metals. The elements do not have to be all metals. For instance, steel is made up of the metal iron and the non-metal carbon.

Nonmetals:

Nonmetals are not as easy to recognize as metals because they do not always share obvious physical properties. However, in general the properties of nonmetals are the opposite of metals. They are dull, brittle, and are not good conductors of heat and electricity.

Nonmetals include solids, gases, and one liquid (bromine).

Nonmetals have four to eight electrons in their outermost energy levels and tend to attract electrons to become anions (negatively charged ions). As a result, the outer levels are usually filled with eight electrons. This difference in the number of electrons is what caused the differences between metals and nonmetals. The outstanding chemical property of nonmetals is that they react with metals.

The **halogens** can be found in Group 17. Halogens combine readily with metals to form salts. Table salt, fluoride toothpaste, and bleach all have an element from the halogen family.

The **Noble Gases** got their name from the fact that they did not react chemically with other elements, much like the nobility did not mix with the masses. These gases (found in Group 18) will only combine with other elements under very specific conditions. They are **inert** (inactive).

In recent years, scientists have found this to be only generally true, since chemists have been able to prepare compounds of krypton and xenon.

Metalloids:

Metalloids have properties in between metals and nonmetals. They can be found in Groups 13 - 16, but do not occupy the entire group. They are arranged in stair steps across the groups.

Physical Properties:
1. All are solids having the appearance of metals.
2. All are white or gray, but not shiny.
3. They will conduct electricity, but not as well as a metal.

Chemical Properties:
1. Have some characteristics of metals and nonmetals.
2. Properties do not follow patterns like metals and nonmetals. Each must be studied individually.

Boron is the first element in Group 13. It is a poor conductor of electricity at low temperatures. However, increase its temperature and it becomes a good conductor. By comparison, metals, which are good conductors, lose their ability as they are heated. It is because of this property that boron is so useful. Boron is a semiconductor. **Semiconductors** are used in electrical devices that have to function at temperatures too high for metals.

Silicon is the second element in Group 14. It is also a semiconductor and is found in great abundance in the earth's crust. Sand is made of a silicon compound, silicon dioxide. Silicon is also used in the manufacture of glass and cement.

Skill 11.6   Interpreting chemical symbols, formulas, and IUPAC nomenclature

The IUPAC is the **International Union of Pure and Applied Chemistry**, an organization that formulates naming rules. **Organic compounds contain carbon**, and they have a separate system of nomenclature but some of the simplest molecules containing carbon also fall within the scope of inorganic chemistry.

Naming rules depend on whether the chemical is an ionic compound or a molecular compound containing only covalent bonds. There are special rules for naming acids. The rules below describe a group of traditional "semi-systematic" names accepted by IUPAC.

Ionic compounds: Cation

Ionic compounds are named with the **cation (positive ion) first**. Nearly all cations in inorganic chemistry are **monatomic**, meaning they just consist of one atom (like Ca^{2+}, the calcium ion.) This atom will be a **metal ion**. For common ionic compounds, the **alkali metals always have a 1+ charge** and the **alkali earth metals always have a 2+ charge.**

Many metals may form cations of more than one charge. In this case, a Roman numeral in parenthesis after the name of the element is used to indicate the ion's charge in a particular compound. This Roman numeral method is known as the **Stock system**. An older nomenclature used the suffix *–ous* for the lower charge and *–ic* for the higher charge and is still used occasionally.

Example: Fe^{2+} is the iron(II) ion, ferrous, and Fe^{3+} is the iron(III) ion, ferric.

The only common inorganic **polyatomic cation** is **ammonium: NH_4^+.**

Ionic compounds: Anion

The **anion** (negative ion) is named and written last. Monatomic anions are formed from nonmetallic elements and are named by **replacing the end of the element's name with the suffix –ide.**

Examples: Cl^- is the chloride ion, S^{2-} is the sulfide ion, and N^{3-} is the nitride ion.

These anions also end with *–ide*:

C_2^{2-}	N_3^-	O_2^{2-}	O_3^-	S_2^{2-}	CN^-	OH^-
carbide or acetylide	azide	peroxide	ozonide	disulfide	cyanide	hydroxide

Oxoanions (also called oxyanions) **contain one element in combination with oxygen.** Many common polyatomic anions are oxoanions that **end with the suffix –ate.** If an element has two possible oxoanions, the one with the element at a lower oxidation state **ends with –ite.** This anion will also usually have **one less oxygen per atom than the –ate ion.** Additional oxoanions are named with the prefix *hypo-* if they have a lower oxidation number (less oxygen) than the *–ite* form and the prefix *per–* if they have a higher oxidation number (more oxygen) than the *–ate* form.

Ionic compounds: Hydrates

Water molecules often occupy positions within the lattice of an ionic crystal. These compounds are called **hydrates**, and the water molecules are known as **water of hydration**. The water of hydration is added after a centered dot in a formula. In a name, a number-prefix (listed below for molecular compounds) indicating the number of water molecules is followed by the root –*hydrate*.

Ionic compounds: Putting it all together

We now have the tools to name most common salts given a formula and to write a formula for them given a name. To determine a formula given a name, the number of anions and cations that are needed to achieve a neutral charge must be found.

Example: Determine the formula of cobalt(II) phosphite octahydrate.

Solution: For the cation, find the symbol for cobalt (Co) and recognize that it is present as Co^{2+} ions from the Roman numerals. For the anion, remember the phosphite ion is PO_3^{3-}. A neutral charge is achieved with 3 Co^{2+} ions for every 2 PO_3^{3-} ions. Add eight (octa = 8) H_2O for water of hydration for the answer:

$$Co_3(PO_3)_2 \cdot 8H_2O.$$

Molecular compounds

Molecular compounds (compounds making up molecules with a neutral charge) are usually composed entirely of nonmetals and are named by placing the **less electronegative atom first**. The suffix –*ide* is added to the second, more electronegative atom, and prefixes indicating numbers are added to one or both names if needed.

Prefix	*mono-*	*di-*	*tri-*	*tetra-*	*penta-*	*hexa-*	*hepta-*	*octa-*	*nona-*	*deca-*
Meaning	1	2	3	4	5	6	7	8	9	10

The final "o" or "a" may be left off these prefixes for oxides.

The electronegativity requirement is the reason the compound with two oxygen atoms and one nitrogen atom is called nitrogen dioxide, NO_2 and <u>not</u> dioxygen nitride O_2N. The hydride of sodium is NaH, sodium hydride, but the hydride of bromine is HBr, hydrogen bromide (or hydrobromic acid if it's in aqueous solution). Oxygen is only named first in compounds with fluorine such as oxygen difluoride, OF_2, and fluorine is never placed first because it is the most electronegative element. Electronegativity decreases to the left and below Fluorine.

<u>Acids</u>

There are special naming rules for acids that correspond with the **suffix of their corresponding anion** if hydrogen were removed from the acid. Normally the cation is H^+.

1. Anions ending with –*ide* correspond to acids with the prefix *hydro*– and the suffix –*ic*. HCl is hydrogen chloride or hydrochloric acid and H_2S is hydrogen sulfide or hydrosulfuric acid.
- Anions ending with –*ate* correspond to acids with no prefix that end with –*ic*. For example, $HClO_3$ is hydrogen chlorate or chloric acid.
- Oxoanions ending with –*ite* have associated acids with no prefix and the suffix –*ous*. $HClO_2$ is hydrogen chlorite or chlorous acid.
- The *hypo*– and *per*– prefixes are maintained. As examples, HClO is hydrogen hypochlorite or hypochlorous acid and $HClO_4$ is hydrogen perchlorate or perchloric acid.

0012   UNDERSTAND CHANGES IN MATTER

Skill 12.1   Analyzing physical, chemical, and nuclear changes in matter and the factors that affect these changes

Matter constantly changes. A **physical change** is a change that does not produce a new substance. The freezing and melting of water is an example of physical change. A **chemical change** (or chemical reaction) is any change of a substance into one or more other substances. Burning materials turn into smoke; a seltzer tablet fizzes into gas bubbles.

Phase changes are good examples of physical changes. The **phase of matter** (solid, liquid, or gas) is identified by its shape and volume. A **solid** has a definite shape and volume. A **liquid** has a definite volume, but no shape. A **gas** has no shape or volume because it will spread out to occupy the entire space of whatever container it is in.

Energy is the ability to cause change in matter. Applying heat energy to a frozen liquid changes it from solid back to liquid. Continue heating it and it will boil and give off steam, a change of some of the liquid to a gas.

Evaporation is the change in phase from liquid to gas. Evaporation requires an input of energy. **Condensation** is the change in phase from gas to liquid. Condensation releases energy. **Sublimation** is the phase change from solid to vapor without going through the liquid phase. **Freezing** releases energy as a substance goes from liquid to solid. **Melting** takes energy to go from solid to liquid.

One or more substances are formed during a **chemical reaction**. Energy is released during some chemical reactions. If energy is released, the reaction is said to be exothermic. Sometimes the energy release is slow and sometimes it is rapid. In a fireworks display, energy is released very rapidly. However, the chemical reaction that produces tarnish on a silver spoon happens very slowly. Other chemical reactions require energy in order to happen. If energy is needed, the reaction is endothermic.

Chemical equilibrium is defined as the rate of forward reaction equals the rate of backward (reverse) reaction. When this occurs, the quantities of reactants and products are at a 'steady state' and no longer shifting.

Synthesis or Combination Reaction: In a synthesis or combination chemical reaction, two elements combine to form a new substance: A + B → C. We can represent the reaction and the results in a chemical equation.
Carbon and oxygen form carbon dioxide. The equation can be written:

$$C \quad + \quad O_2 \quad \rightarrow \quad CO_2$$

$$\text{1 atom of} \quad + \quad \text{2 atoms of} \quad \rightarrow \quad \text{1 molecule of}$$
$$\text{carbon} \qquad \qquad \text{oxygen} \quad \rightarrow \quad \text{carbon dioxide}$$

No matter is ever gained or lost during a chemical reaction according to the **law of conservation of mass**; therefore the chemical equation must be *balanced*. This means that there must be the same number of atoms of each element on both sides of the equation. Remember that the subscript numbers indicate the number of atoms of the elements. If there is no subscript, assume there is only one atom.

Decomposition Reaction: In a decomposition chemical reaction, the molecules of a substance split forming two or more new substances: C → A + B. An electric current can split water molecules into hydrogen and oxygen gas.

$$2H_2O \quad \rightarrow \quad 2H_2 \quad + \quad O_2$$

$$\text{2 molecules} \quad \rightarrow \quad \text{2 molecules} \quad + \quad \text{1 molecule}$$
$$\text{of water} \qquad \qquad \text{of hydrogen} \qquad \qquad \text{of oxygen}$$

The number of molecules is shown by the number in front of an element or compound. If no number appears, assume that it is 1 molecule.

Single Replacement Reaction: A single replacement chemical reaction is when elements change places with each other: AB + C → CB + A. An example of one element taking the place of another is when iron becomes a cation and changes places with copper in the compound copper sulfate forcing copper out:

$CuSo_4$	+	Fe	→	$FeSO_4$ + Cu
copper	+	iron		iron copper
sulfate		(steel wool)		sulfate

Double Displacement or Double Replacement Reaction: Sometimes two sets of elements change places: AB + CD → AD + CB. In this example, an acid and a base are combined to yield a salt and water:

HCl	+	NaOH	→	NaCl	+	H_2O
hydrochloric acid		sodium hydroxide		sodium chloride (table salt)		water

Combustion Reaction: Many substances, especially organic substances, can be burned in the presence of oxygen: $CH_x + O_2 \rightarrow yCO_2 + zH_2O$. This produces carbon dioxide and water:

CH_4	+	$2 O_2$	→	CO_2	+	$2 H_2O$
Methane		oxygen		carbon dioxide		water

Matter can change, but it can not be created or destroyed. The sample equations show two things:

1. In a chemical reaction, matter is changed into one or more different kinds of matter.
2. The amount of matter present before and after the chemical reaction is the same.

Many chemical reactions give off energy. Like matter, energy can change form but it can be neither created nor destroyed during a chemical reaction. This is the **law of conservation of energy.**

Skill 12.2   Recognizing types, characteristics, and applications of radioactivity and radioactive decay

Radioactivity is the breaking down of atomic nuclei by releasing particles or electromagnetic radiation. Radioactive nuclei give off radiation in the form of streams of particles or energy. Alpha particles are positively charged particles consisting of two protons and two neutrons. It is the slowest form of radiation. It can be stopped by paper! Beta particles are electrons. It is produced when a neutron in the nucleus breaks up into a proton and an electron. The proton remains inside the nucleus, increasing its atomic number by one. But the electron is given off. They can be stopped by aluminum. Gamma rays are electromagnetic waves with extremely short wavelengths. They have no mass. They have no charge so they are not deflected by an electric field. Gamma rays travel at the speed of light. It takes a thick block of lead to stop them. Uranium is the source of radiation and therefore is radioactive. Marie Curie discovered new elements called radium and polonium that actually give off more radiation than uranium.

The major concern with radioactivity is in the case of a nuclear disaster. Medical misuse is also a threat. Radioactivity ionizes the air it travels through. It is strong enough to kill cancer cells or dangerous enough to cause illness or even death. Gamma rays can penetrate the body and damage cells. Protective clothing is needed when working with gamma rays. Electricity from nuclear energy uses uranium 235. The devastation of the Russian nuclear power plant disaster has evacuated entire regions as the damage to the land and food source will last for hundreds of years.

Isotopes are called radioisotopes when they have unstable nuclei that are radioactive. **Alpha particles** (α) are positively charged particles ($^+2$) emitted from a radioactive nucleus. They consist of two protons and two neutrons and are identical to the nucleus of a helium atom ($^4_2$He).

Example: $^{238}_{92}\text{U} \rightarrow {}^4_2\text{He} + {}^{234}_{90}\text{Th}$

When an atom loses an alpha particle, the Z number (atomic number) is lower by two, so move back two spaces on the periodic table to find what the new element is. The new element has an A number (atomic mass number) that is four less than the original element. Because alpha particles are large and heavy, paper or clothing or even dead skin cells shield from their effects.

Beta rays (β) are negatively charged (-1) and fast moving because they are actually electrons. They are written as an electron $^0_{-1}$ e (along with a proton) which is emitted from the nucleus as a neutron decays. Carbon-14 decays by emitting a beta particle.

Example: $^{14}_6\text{C} \rightarrow {}^{14}_7\text{N} + {}^0_{-1}\text{e}$

The Z number actually adds one since its total must be the same on both the left and the right of the arrow and the electron on the right adds a negative one. The A number is unchanged. The Z determines what the element is, so look for it on the periodic table to determine the product. Metal foil or wood is needed to shield from its effects.

Gamma rays (γ) are high energy electromagnetic waves. They are the same kind of radiation as visible light but of much shorter wavelength and higher frequency. Gamma rays have no mass or charge, so the Z and A numbers are not affected. Radioactive atoms often emit gamma rays along with either alpha or beta particles. Protection from gamma radiation takes lead or concrete.

$$\text{Example 1:} \quad ^{226}_{88}\text{Ra} \rightarrow \,^{222}_{86}\text{Rn} + \,^{4}_{2}\text{He} + \gamma$$

$$\text{Example 2:} \quad ^{234}_{90}\text{Th} \rightarrow \,^{234}_{91}\text{Pa} + \,^{0}_{-1}\text{e} + \gamma$$

A **positron** is a particle with the mass of an electron but a positive charge ($^{0}_{+1}\text{e}$). It may be emitted as a proton changes to a neutron.

Transmutation is the conversion of an atom of one element to an atom of another element such as occurs in alpha and beta radiation. It also occurs when high-energy particles (such as protons, neutrons, or alpha particles) bombard the nucleus of an atom. The elements in the periodic table with atomic numbers above 92 are called the transuranium elements, all of which are radioactive elements that have been synthesized in nuclear reactors and nuclear accelerators.

$$\text{Example:} \quad ^{238}_{92}\text{U} + \,^{1}_{0}n \rightarrow \,^{239}_{92}\text{U} \rightarrow \,^{0}_{-1}\text{e} + \,^{239}_{93}\text{Np} \rightarrow \,^{239}_{94}\text{Pu} + \,^{0}_{-1}\text{e}$$

Nuclear fission is the splitting of a nucleus into smaller fragments by bombardment with neutrons. Fission releases enormous amounts of energy. Controlled fission is the source of the energy in nuclear power plants.

In **nuclear fusion** hydrogen nuclei fuse to make helium nuclei. Nuclear fusion releases even more energy than fission. The main disadvantage of using nuclear fusion for generating electricity is that it requires very high temperatures and a very large activation energy, neither of which is practical. The sun's energy is generated by nuclear fusion, i.e. combination of smaller nuclei into a larger nucleus. Fusion creates much less radioactive waste,

Every radioisotope has its own characteristic rate of decay. The **half-life of an isotope** is the time it takes for half the original amount of the isotope in a given sample to decay. For example, the half-life of carbon-14 is 5700 years. If there are 25 grams of carbon-14 in a petrified log, then 5700 years later it will contain 12.5 grams of carbon-14. Another 5700 years later it will contain 6.25 grams of C-14.

Skill 12.3 **Applying knowledge of the law of conservation of matter to chemical equations**

Matter can change, but it can not be created or destroyed. This is a restatement of the Law of Conservation.

Please see Section 12.1 for discussion of types of chemical equations and specific examples of how matter is conserved.

Balancing Chemical Equations

1. Determine the correct formulas for all reactants and products, using subscripts to balance ionic charges.
2. Write formulas for reactants on the left of the arrow and predict the products and write their formulas to the right of the arrow.
3. Under the reactants list all the elements in the reactants, starting with metals, then nonmetals, listing oxygen last and hydrogen next to last. Under the products, list all the elements in the same order as those under the reactants (straight across from them).
4. Count the atoms of each element on the left side and list the numbers next to the elements. Repeat for products. Don't forget that subscripts outside a parenthesis multiply everything inside the parenthesis including subscripts inside the parenthesis.
5. For the first element in the list that has unequal numbers of atoms, use a coefficient (numeral to the left of the compound or element) to give the correct number of atoms. NEVER change the subscripts to balance an equation.
6. Go to the next unbalanced element and balance it, moving down the list until all are balanced.
7. Start back at the beginning of the list and actually count the atoms of each element on each side of the arrow to make sure the number listed is the actual number. Re-balance and re-check as needed.

Example: $Al(OH)_3 + NaOH \rightarrow NaAlO_2 + 2\,H_2O$

Al = 1	Al = 1
Na = 1	Na = 1
H = 3 + 1 = 4	H = ~~2~~ 4
O = 3 + 1 = 4	O = ~~2 + 1 =~~ 3 2 + 2 = 4
	Put a 2 in front of H_2O

Skill 12.4 **Identifying the components and properties of solutions, including acids and bases, and factors that affect solubility**

When two or more pure materials mix in a homogeneous way (with their molecules intermixing on a molecular level), the mixture is called a **solution**.

When a pure liquid and a gas or solid form a liquid solution, the pure liquid is called the **solvent** and the non-liquids are called **solutes**. When all components in the solution were originally liquids, then the one present in the greatest amount is called the solvent and the others are called solutes. Liquids that are soluble in one another are miscible. A solution at equilibrium with undissolved solute is a **saturated** solution. The amount of solute required to form a saturated solution in a given amount of solvent is called the **solubility** of that solute. If less solute is present, the solution is called **unsaturated**. It is also possible to have more solute than the equilibrium amount, resulting in a solution that is termed **supersaturated**.

Aqueous solutions are water samples containing dissolved substances. The dissolving medium (water) is called the **solvent** and the dissolved particles of a substance are called the **solute**. The **rate of solubility** can be increased by (1) increasing the temperature, (2) agitating by stirring or shaking the solution, or (3) decreasing the solute particle size to create more surface area.

Water molecules are dipoles, so molecules which are dipoles dissolve well in it. With all the molecules in constant motion, it is easy for the H^+ end of H_2O to attract the negative ions and the O^{-2} end of H_2O to attract the positive ions, thus breaking the bonds between positive and negative ions of the solute. **Solvation** occurs when those ions are surrounded by water molecules.

Sometimes ionic compounds have stronger attractive forces than the attractive forces exerted by water. Those compounds, like barium sulfate and calcium carbonate, are insoluble ionic compounds.

Nonpolar molecules do not dissolve much in water, but must be dissolved in a nonpolar solvent such as benzene. As a rule, "like dissolves like" holds true since polar solvents like water dissolve polar and ionic compounds and nonpolar solvents dissolve nonpolar and organic compounds.

An **acid** often contains hydrogen (H). Although it is never wise to taste a substance to identify it, acids have a sour taste. Vinegar and lemon juice are both acids, and acids occur in many foods in a weak state. Strong acids can burn skin and destroy materials. Common acids include:

Sulfuric acid (H_2SO_4)	-	Used in medicines, alcohol, dyes, and car batteries.
Nitric acid (HNO_3)	-	Used in fertilizers, explosives, cleaning materials.
Carbonic acid (H_2CO_3)	-	Used in soft drinks.
Acetic acid ($HC_2H_3O_2$)	-	Used in making plastics, rubber, photographic film, and as a solvent.

Bases have a bitter taste and the stronger ones feel slippery. Like acids, strong bases can be dangerous and should be handled carefully. Many bases contain the elements oxygen and hydrogen (OH). Many household cleaning products contain bases. Common bases include:

Sodium hydroxide	NaOH	-	Used in making soap, paper, vegetable oils, and refining petroleum.
Ammonium hydroxide	NH_4OH	-	Making deodorants, bleaching compounds, cleaning compounds.
Potassium hydroxide	KOH	-	Making soaps, drugs, dyes, alkaline batteries, and purifying industrial gases.
Calcium hydroxide	$Ca(OH)_2$	-	Making cement and plaster

Historically there are three definitions of acids and bases:

The **Lewis** definition is the broadest. According to Lewis an acid is an electron pair acceptor and a base is an electron pair donor. The acid becomes more negative as it goes through the reaction and the base becomes more positive.

Arrhenius' definition said that all bases end in OH and all acids start with H. Thus, the acid donates H and the base donates OH to create a water molecule and the remaining cation and anion combine to create a salt.

The **Bronsted-Lowry** definition calls an acid a H+ donor and the base an H+ acceptor. Rather than looking at individual compounds, one must look at the full chemical reaction equation to determine which compound on the left can accept an H+ to become a compound on the right. This pair becomes the conjugate base-acid pair. The other pair should be a compound on the left that can donate an H+ to become the remaining compound on the right. That makes them the conjugate acid-base pair. An **indicator** is a substance that changes color when it comes in contact with an acid or a base. Litmus paper is an indicator. Blue litmus paper turns red in an acid. Red litmus paper turns blue in a base.

A substance that is neither acid nor base is **neutral**. Neutral substances do not change the color of litmus paper.

A **salt** is formed when an acid and a base combine chemically. Water is also formed. The process is called **neutralization**. Table salt (NaCl) is an example of this process. Salts are also used in toothpaste, epsom salts, and cream of tartar. Calcium chloride ($CaCl_2$) is used on frozen streets and walkways to melt the ice.

Skill 12.5   Analyzing factors that affect rates of chemical reactions (e.g., temperature, catalysts)

Kinetic energy is the energy of motion. Kinetic molecular theory says the particles of matter are in a state of constant motion. The particles of matter do not lose energy in collisions. Increasing the temperature causes the molecules to move more rapidly and in a greater area. The rate of most simple reactions **increases with temperature** because a **greater fraction of molecules have the kinetic energy** required to overcome the reaction's activation energy.

Activation energy is the energy required to loosen bonds in molecules to allow them to become reactive. The chart below shows the effect of temperature on the distribution of kinetic energies in a sample of molecules. These curves are called **Maxwell-Boltzmann distributions**. The shaded areas represent the fraction of molecules containing sufficient kinetic energy for a reaction to occur. This area is larger at a higher temperature; so more molecules are above the activation energy and more molecules react per second.

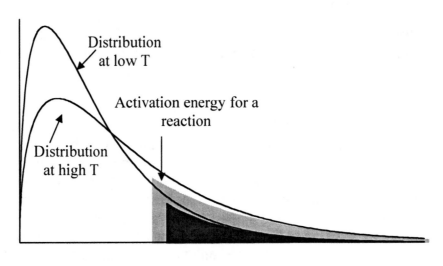

http://www.mhhe.com/physsci/chemistry/essentialchemistry/flash/activa2.swf
provides an animated audio tutorial on energy diagrams.

Kinetic molecular theory may be applied to reaction rates in addition to physical constants like pressure. **Reaction rates increase with reactant concentration** because more reactant molecules are present and more are likely to collide with one another in a certain volume at higher concentrations. The nature of these relationships determines the rate law for the reaction. For ideal gases, the concentration of a reactant is its molar density, and this varies with pressure and temperature as discussed in.

Kinetic molecular theory also predicts that **reaction rate constants (values for k) increase with temperature** for one of two reasons:

1. More reactant molecules will collide with each other per second.
2. These collisions will each occur at a higher energy that is more likely to overcome the activation energy of the reaction.

A **catalyst** is a material that increases the rate of a chemical reaction without being changed permanently itself in the process. Catalysts provide an alternate reaction mechanism for the reaction to proceed in the forward and sometimes in the reverse direction. Therefore, **catalysts have no impact on the chemical equilibrium** of a reaction. They will not make a less favorable reaction more favorable.

Catalysts reduce the activation energy of a reaction. This is the amount of energy needed for the reaction to begin. Molecules with such low energies that they would have taken a long time to react will react more rapidly if a catalyst is present.

The impact of a catalyst may also be represented on an energy diagram. **A catalyst increases the rate of both the forward and reverse reactions by lowering the activation energy** for the reaction. Catalysts provide a different activated complex for the reaction at a lower energy state.

Biological catalysts are called **enzymes.**

Skill 12.6 Identifying characteristics of a system at equilibrium

Several factors are important in the progress of a reaction. The most basic first step in any reaction is that the molecules of the reactants must collide with one another. However, only a fraction of the collisions between the reactants allow the reaction to begin. This is because the molecules must collide in the proper orientation and with sufficiently high energy. This **activation energy E_a** is the minimum energy needed to overcome the barrier to the formation of products. That is, it is the minimum energy needed for the reaction to occur. The activation energy, E_a, is the difference between the energy of reactants and the energy of the activated complex, which is an intermediate form.

The energy change during the reaction, ΔE, is the difference between the energy of the products and the energy of the reactants. If the energy of the products is lower than that of the reactants, the reaction will be **exothermic**. If reactants are lower than products, the reaction will be **endothermic**. Thus, energy must be added to allow an endothermic reaction to progress towards equilibrium. However, high activation energies can be overcome with the use of a catalyst. A catalyst decreases E_a and so increases the rate of both the forward and reverse reactions by lowering the activation energy for the reaction. Note that a catalyst does not change the position of equilibrium, but merely reduces the energy requirements of a reaction.

If we return to the question of collisions between molecules, it becomes apparent that kinetic energy will have a bearing on the likelihood of these collisions. Molecules that are moving around quickly will be more likely to collide and the higher number of collisions also increases the chance that the molecules will meet in the correct orientation. Additionally, if the molecules generally have higher energy, it will be more likely for them to obtain the E_a required for the reaction to proceed.

Many reactions occur with the reactants going completely to products and then the reaction is over (A + B → AB). However, a lot of reactions start with reactants going to products (A + B → AB) and then some of the products break down into reactants (A + B ← AB) while more reactants become products until **equilibrium** of the reaction is attained and maintained. Equilibrium is when the amount of reactants becoming products is equal to the amount of products becoming reactants (A + B ↔ AB).

According to **Le Châtelier's Principle** if a stress is applied to a system in a dynamic equilibrium, the system changes to relieve the stress. Stresses that disturb equilibrium are:

1. Change in concentration – if a reactant's concentration is increased, the equilibrium is displaced to the right ($\rightarrow$) meaning that the reactants are used up faster, more products are formed ($\leftarrow$), and the new equilibrium has a lower concentration of reactants. Conversely, an increase in a product's concentration displaces the reaction equilibrium to the left, favoring the reactants.

2. Change in pressure – this only applies to gases where an increase in pressure displaces the reaction equilibrium to the right ($\rightarrow$)

3. Change in temperature – addition of heat favors the endothermic reaction; however, a rise in temperature increases the rate of any reaction.

Therefore, anything that increases the probability of these collisions and the energy of the molecules will speed the reaction's obtainment of equilibrium. Thus it is clear that temperature, pressure, and concentration must have an effect on systems in equilibrium. Their effect is generalized in **Le Châtelier's Principle**.

0013 UNDERSTAND PRINCIPLES AND CONCEPTS RELATED TO ENERGY

Skill 13.1 Identifying forms (e.g., mechanical, chemical) and types (e.g., potential, kinetic) of energy and their characteristics

Abstract concept it might be, but energy is one of the most fundamental concepts in our world. We use it to move people and things from place to place, to heat and light our homes, to entertain us, to produce food and goods and to communicate with each other. It is not some sort of magical invisible fluid, poured, weighed or bottled. It is not a substance but rather the ability possessed by things.

Technically, **energy is the ability to do work or supply heat.** Work is the transfer of energy to move an object a certain distance. It is the motion against an opposing force. Lifting a chair into the air is work; the opposing force is gravity. Pushing a chair across the floor is work; the opposing force is friction.

Heat, on the other hand, is <u>not a form</u> of energy but a <u>method of transferring</u> energy.

This energy, according to the **First Law of Thermodynamics**, is conserved. That means energy is neither created nor destroyed in ordinary physical and chemical processes (non-nuclear). Energy is merely changed from one form to another. Energy in all of its forms must be conserved. In any system, $\Delta E = q + w$ (E = energy, q = heat and w = work).

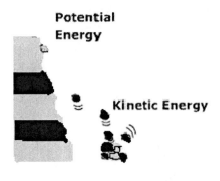

Potential Energy

Kinetic Energy

Energy exists in two basic forms, potential and kinetic. **Kinetic energy** is the energy of a moving object. **Potential energy** is the energy stored in matter due to position relative to other objects.

In any object, solid, liquid or gas, the atoms and molecules that make up the object are constantly moving (vibrational, translation and rotational motion) and colliding with each other. They are not stationary.

Due to this motion, the object's particles have varying amounts of kinetic energy. A fast moving atom can push a slower moving atom during a collision, so it has energy. All moving objects have energy and that energy depends on the object's mass and velocity. Kinetic energy is calculated:

$$KE = \tfrac{1}{2} mv^2$$

Where KE = kinetic energy; m = mass of object; v = velocity

The temperature exhibited by an object is proportional to the average kinetic energy of the particles in the substance. Increase the temperature of a substance and its particles move faster so their average kinetic energies increase as well. But temperature is NOT energy; it is not conserved.

The energy an object has due to its position or arrangement of its parts is called potential energy. Potential energy due to position is equal to the mass of the object times the gravitational pull on the object times the height of the object, or:

$$PE = mgh$$

Where PE = potential energy; m = mass of object; g = gravity; and h = height.

Heat is energy that is transferred between objects caused by differences in their temperatures. Heat passes spontaneously from an object of higher temperature to one of lower temperature. This transfer continues until both objects reach the same temperature. Both kinetic energy and potential energy can be transformed into heat energy. When you step on the brakes in your car, the kinetic energy of the car is changed to heat energy by friction between the brake and the wheels. Other transformations can occur from kinetic to potential as well. Since most of the energy in our world is in a form that is not easily used, man and mother nature have developed some clever ways of changing one form of energy into another form that may be more useful.

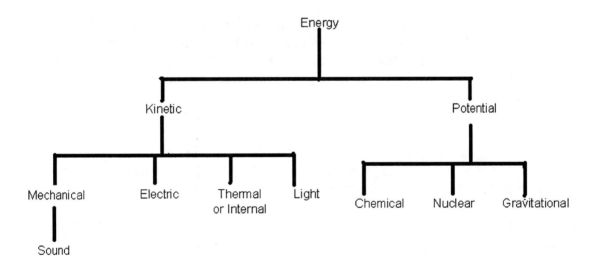

Gravitational Potential Energy:

When something is lifted or suspended in air, work is done on the object against the pull of gravity. This work is converted to a form of potential energy called gravitational potential energy.

Nuclear Potential Energy:

The nuclear energy trapped inside the atom is referred to as nuclear energy. When the atom is split, tremendous energy is released in the form of heat and light.

Chemical Potential Energy:

The energy generated from chemical reactions in which the chemical bonds of a substance are broken and rearranged to form new substances is called chemical potential energy.

Electrical Kinetic energy:

The flow of electrons along a circuit is called electrical energy. The movement of electrons creates an electric current which generates electricity

Mechanical Kinetic Energy:

Mechanical energy is the energy of motion doing work, like a pendulum moving back and forth in a grandfather clock.

Thermal Kinetic Energy:

Thermal Energy is defined as the energy that a substance has due to the chaotic motion of its molecules. Molecules are in constant motion, and always possess some amount of kinetic energy. It is also called Internal energy and is not the same as heat.

Light or Radiant Kinetic Energy:

Radiant energy comes from a light source, such as the sun. Energy released from the sun is in the form of photons. These tiny particles, invisible to the human eye, move in a way similar to a wave.

Energy transformations make it possible for us to use energy; do work. Here are some examples of how energy is transformed to do work:

1. Different types of stoves are used to transform the chemical energy of the fuel (gas, coal, wood, etc.) into heat.

2. Solar collectors can be used to transform solar energy into electrical energy.

3. Windmills make use of the kinetic energy of the air molecules, transforming it into mechanical or electrical energy.

4. Hydroelectric plants transform the kinetic energy of falling water into electrical energy.

5. A flashlight converts chemical energy stored in batteries to light energy and heat. Most of the energy is converted to heat, only a small amount is actually changed into light energy.

Skill 13.2   Applying knowledge of the law of conservation of energy in the analysis of physical and chemical systems

The relationship between heat, forms of energy, and work (mechanical, electrical, etc.) are the **Laws of Thermodynamics.** These laws deal strictly with systems in thermal equilibrium and not those within the process of rapid change or in a state of transition. Systems that are nearly always in a state of equilibrium are called **reversible systems.**

The first law of thermodynamics is a restatement of conservation of energy. The change in heat energy supplied to a system (Q) is equal to the sum of the change in the internal energy (U) and the change in the work done by the system against internal forces. $\Delta Q = \Delta U + \Delta W$

The second law of thermodynamics is stated in two parts:

1. No machine is 100% efficient. It is impossible to construct a machine that only absorbs heat from a heat source and performs an equal amount of work because some heat will always be lost to the environment.

2. Heat can not spontaneously pass from a colder to a hotter object. An ice cube sitting on a hot sidewalk will melt into a little puddle, but it will never spontaneously cool and form the same ice cube. Certain events have a preferred direction called the **arrow of time.**

Entropy is the measure of how much energy or heat is available for work. Work occurs only when heat is transferred from hotter to cooler objects. Once this is done, no more work can be extracted. The energy is still being conserved, but is not available for work as long as the objects are at the same temperature. Theory has it that, eventually, all things in the universe will reach the same temperature. If this happens, energy will no longer be usable.

Entropy may be thought of as **the disorder in a system** or as a measure of the **number of states a system may occupy**. Changes due to entropy occur in one direction with no driving force. For example, a small volume of gas released into a large container will expand to fill it, but the gas in a large container never spontaneously collects itself into a small volume. This occurs because a large volume of gas has more disorder and has more places for gas molecules to be.

This change occurs because **processes increase in entropy** when given the opportunity to do so. Entropy has units of Joules per Kelvin (J/K).

If two different chemicals are at the same temperature, in the same state of matter, and they have the same number of molecules, their entropy difference will depend mostly on the number of ways the atoms within the two chemicals can rotate, vibrate, and flex. Most of the time, **the more complex molecule will have the greater entropy** because there are more energetic and spatial states in which it may exist.

The **entropy change of a reaction**, ΔS, is given by the sum of the absolute entropies of all the products multiplied by their stoichiometric coefficients minus the sum of all the products multiplied by their stoichiometric coefficients:

The **enthalpy** (*H*) of a material is the **sum of its internal energy and the mechanical work** it can do by driving a piston. We usually don't deal with mechanical work in high school chemistry, so the differences between internal energy and enthalpy are not important.

When a chemical reaction takes place, the enthalpies of the products will differ from the enthalpies of the reactants. There is an energy change for the reaction ΔH_{rxn}, determined by **the sum of the enthalpies of the products minus the sum of the enthalpies of the reactants**:

$$\Delta H_{rxn} = H_{product\,1} + H_{product\,2} + \ldots - \left(H_{reactant\,1} + H_{reactant\,2} + \ldots\right).$$

The enthalpy change for a reaction is commonly called the **heat of reaction**.

If the sum of the enthalpies of the products is greater than the sum of the enthalpies of the reactants, then ΔH_{rxn} **is positive** and the reaction is **endothermic**. Endothermic reactions **absorb heat** from their surroundings. The simplest endothermic reactions break chemical bonds.

If the sum of the enthalpies of the products is less than the sum of the enthalpies of the reactants, then ΔH_{rxn} **is negative** and the reaction is **exothermic**. Exothermic reactions **release heat** into their surroundings.

The enthalpy change of a reaction ΔH_{rxn} **is equal in magnitude but has the opposite sign to the enthalpy change for the reverse reaction**. If a series of reactions lead back to the initial reactants then the net energy change for the entire process is zero.

When a reaction is composed of sub-steps, the total enthalpy change will be the sum of the changes for each step. The ability to add together these enthalpies to form ultimate products from initial reactants is known as **Hess's Law.**

A reaction with a **negative ΔH and a positive ΔS** causes a decrease in energy and an increase in entropy. **These reactions will always occur spontaneously**. A reaction with a positive ΔH and a negative ΔS causes an increase in energy and a decrease in entropy. These reactions never occur to an appreciable extent because the reverse reaction takes place spontaneously.

Whether reactions with the remaining two possible combinations (ΔH and ΔS both positive or both negative) occur depends on the temperature. If **ΔH–TΔS** (known as the **Gibbs Free Energy,** ΔG) is **negative, the reaction will take place**. If it is positive, the reaction will not occur to an appreciable extent. If ΔH–TΔS = 0 exactly, then at equilibrium there will be 50% reactants and 50% products.

A spontaneous reaction is called *exergonic*. A non-spontaneous reaction is known as *endergonic*. These terms are used much less often than *exothermic* and *endothermic*.

Skill 13.3   Demonstrating knowledge of energy transformations and transfers (e.g., heat transfer, energy conversion) in a system

Heat energy that is transferred into or out of a system is **heat transfer.** The temperature change is positive for a gain in heat energy and negative when heat is removed from the object or system.

The formula for heat transfer is $Q = mc\Delta T$ where Q is the amount of heat energy transferred, m is the mass of substance (in kilograms), c is the specific heat of the substance, and ΔT is the change in temperature of the substance. It is important to assume that the objects in thermal contact are isolated and insulated from their surroundings.
If a substance in a closed container loses heat, then another substance in the container must gain heat.

A **calorimeter** uses the transfer of heat from one substance to another to determine the specific heat of the substance. One of the most familiar methods of measuring a quantity of heat is by imparting this heat to a known mass of water and observing the change it produces in the temperature of the water.

<u>Example</u>: *A 0.450-kg cylinder of lead is heated to 100 °C and then dropped into a 50-g copper calorimeter containing 0.100 kg of water at 10 °C. The water is stirred until equilibrium is established, at which time the temperature of the whole system is 21.1 °C. Find the specific heat of lead.*

Heat gained by the water + Heat gained by the copper = heat lost by the lead
$$(m\ C_p\ \Delta T)_{water} + (m\ C_p\ \Delta T)_{copper} = (m\ C_p\ \Delta T)_{lead}$$

$$0.100\ kg \times 1\ kcal/(kg\text{-}C°)(21.1 - 10\ C°) + 0.050\ kg \times 0.093\ kcal/(kg\text{-}C°)(21.1 - 10\ C°) = 0.450\ kg \times (c)(100 - 21.1\ C°)$$
$$1.16 = 35.5\ (c)$$
$$0.033\ kcal/(kg)(C°) = c$$

When an object undergoes a change of phase it goes from one physical state (solid, liquid, or gas) to another. For instance, water can go from liquid to solid (freezing) or from liquid to gas (boiling). The heat that is required to change from one state to the other is called **latent heat.**

The **heat of fusion** is the amount of heat that it takes to change from a solid to a liquid or the amount of heat released during the change from liquid to solid.

The **heat of vaporization** is the amount of heat that it takes to change from a liquid to a gaseous state.

Heat is transferred in three ways: **conduction, convection, and radiation.**

Conduction occurs when heat travels through the heated solid. The transfer rate is the ratio of the amount of heat per amount of time it takes to transfer heat from area of an object to another. For example, if you place an iron pan on a flame, the handle will eventually become hot. How fast the handle gets too hot to handle is a function of the amount of heat and how long it is applied. Because the change in time is in the denominator of the function, the shorter the amount of time it takes to heat the handle, the greater the transfer rate.

Convection is heat transported by the movement of a heated substance. Warmed air rising from a heat source such as a fire or electric heater is a common example of convection. Convection ovens make use of circulating air to more efficiently cook food.

Radiation is heat transfer as the result of electromagnetic waves. The sun warms the earth by emitting radiant energy.

An example of all three methods of heat transfer occurs in the thermos bottle or Dewar flask. The bottle is constructed of double walls of Pyrex glass that have a space in between. Air is evacuated from the space between the walls and the inner wall is silvered. The lack of air between the walls lessens heat loss by convection and conduction. The heat inside is reflected by the silver, cutting down heat transfer by radiation. Hot liquids remain hotter and cold liquids remain colder for longer periods of time.

Skill 13.4 Applying knowledge of the kinetic molecular theory in the analysis of the properties and behavior of solids, liquids, gases, and plasmas

Gas **pressure** results from molecular collisions with container walls. The **number of molecules** striking an **area** on the walls and the **average kinetic energy** per molecule are the only factors that contribute to pressure. A higher **temperature** increases speed and kinetic energy. There are more collisions at higher temperatures, but the average distance between molecules does not change, and thus density does not change in a sealed container.

Kinetic molecular theory (KMT) explains how the pressure and temperature influences behavior of gases by making a few assumptions, namely:

1) The energies of intermolecular attractive and repulsive forces may be neglected.
2) The average kinetic energy of the molecules is proportional to absolute temperature.
3) Energy can be transferred between molecules during collisions and the collisions are elastic, so the average kinetic energy of the molecules doesn't change due to collisions.
4) The volume of all molecules in a gas is negligible compared to the total volume of the container.

Strictly speaking, molecules also contain some kinetic energy by rotating or experiencing other motions. The motion of a molecule from one place to another is called **translation**. Translational kinetic energy is the form that is transferred by collisions, and kinetic molecular theory ignores other forms of kinetic energy because they are not proportional to temperature.

Molecules have **kinetic energy** (they move around), and they also have **intermolecular attractive forces** (they stick to each other). The relationship between these two determines whether a collection of molecules will be a gas, liquid, or solid.

A **gas** has an indefinite shape and an indefinite volume. The kinetic model for a gas is a collection of widely separated molecules, each moving in a random and free fashion, with negligible attractive or repulsive forces between them. Gases will expand to occupy a larger container so there is more space between the molecules. Gases can also be compressed to fit into a small container so the molecules are less separated. **Diffusion** occurs when one material spreads into or through another. Gases diffuse rapidly and move from one place to another.

A **liquid** has a specific volume. The kinetic model for a liquid is a collection of molecules attracted to each other with sufficient strength to keep them close to each other but with insufficient strength to prevent them from moving around randomly. Liquids have a higher density and are much less compressible than gases because the molecules in a liquid are closer together. Diffusion occurs more slowly in liquids than in gases because the molecules in a liquid stick to each other and are not completely free to move.

A **solid** has a definite volume and definite shape. The kinetic model for a solid is a collection of molecules attracted to each other with sufficient strength to essentially lock them in place. Each molecule may vibrate, but it has an average position relative to its neighbors. If these positions form an ordered pattern, the solid is called **crystalline**. Otherwise, it is called **amorphous**. Solids have a high density and are almost incompressible because the molecules are close together. Diffusion occurs extremely slowly because the molecules almost never alter their position.

In a solid, the energy of intermolecular attractive forces is much stronger than the kinetic energy of the molecules, so kinetic energy and kinetic molecular theory are not very important. As temperature increases in a solid, the vibrations of individual molecules grow more intense and the molecules spread slightly further apart, decreasing the density of the solid.

In a liquid, the energy of intermolecular attractive forces is about as strong as the kinetic energy of the molecules and both play a role in the properties of liquids.

In a gas, the energy of intermolecular forces is much weaker than the kinetic energy of the molecules. Kinetic molecular theory is usually applied for gases and is best applied by imagining ourselves shrinking down to become a molecule and picturing what happens when we bump into other molecules and into container walls.

Skill 13.5   Applying knowledge of the gas laws (e.g., Boyle's law, Charles's law)

As a substance is heated, the molecules begin moving faster within the container. As the substance becomes a gas and those molecules hit the sides of the container, pressure builds. **Pressure** is the force exerted on each unit of area of a surface. Pressure is measured in a unit called the **Pascal**. One Pascal (pa) is equal to one Newton of force pushing on one square meter of area.

Volume, temperature, and pressure of a gas are related. Temperature is measured on the Kelvin scale where K = °C + 273.

Temperature and pressure: As the temperature of a gas increases, its pressure increases. When you drive a car, the friction between the road and the tire heats up the air inside the tire. Because the temperature increases, so does the pressure of the air on the inside of the tire. **Guy-Lussac's Law**:

$$T_1 P_2 - T_2 P_1$$

Temperature and Volume: At a constant pressure, an increase in temperature causes an increase in the volume of a gas. If you apply heat to an enclosed container of gas, the pressure inside the bottle will increase as the heat increases. This is called **Charles' Law**.

$$T_1 V_2 = T_2 V_1$$

These relations (pressure and temperature, and temperature and volume) are **direct variations**. As one component increases (decreases), the other also increases (decreases).

However, pressure and volume vary inversely.

Pressure and volume: At a constant temperature, a decrease in the volume of a gas causes an increase in its pressure. An example of this is a tire pump. The gas pressure inside the pump gets bigger as you press down on the pump handle because you are compressing the gas, or forcing it to exist in a smaller volume. This relationship between pressure and volume is called **Boyle's Law**:

$$P_1 V_1 = P_2 V_2$$

Therefore, the Combined Gas Law is:

$$T_1 P_2 V_2 = T_2 P_1 V_1$$

With the addition of the concept of Avogadro's Law, the Ideal Gas Law was developed:

$$P V = n R T$$

Where P = the absolute pressure (Pascals); V = volume (cubic meters); n = number of moles of gas; R = ideal gas constant (8.3145 J/(mol K); T = temperature (Kelvin)

Skill 13.6 Analyzing phase diagrams (e.g., heat versus temperature) and the flow of energy during changes in states of matter

Heat and temperature are different physical quantities. **Heat** is a measure of energy. **Temperature** is the measure of how hot (or cold) a body is with respect to a standard object.

Two concepts are important in the discussion of temperature changes. Objects are in thermal contact if they can affect each other's temperatures. Set a hot cup of coffee on a desk top. The two objects are in thermal contact with each other and will begin affecting each other's temperatures. The coffee will become cooler and the desktop warmer. Eventually, they will have the same temperature. When this happens, they are in **thermal equilibrium.**

We can not rely on our sense of touch to determine temperature because the heat from a hand may be conducted more efficiently by certain objects, making them feel colder. **Thermometers** are used to measure temperature. A small amount of mercury in a capillary tube will expand when heated. The thermometer and the object whose temperature it is measuring are put in contact long enough for them to reach thermal equilibrium. Then the temperature can be read from the thermometer scale.

If a time graph was made of a pure substance being heated or cooled, it would look something like this graph for the heating of water. Different changes are taking place during each interval on the graph.

When the system is heated, energy is transferred into it. In response to the energy it receives, the system changes, either by increasing its temperature or changing phase.

During the interval marked A on the graph below, energy is being absorbed by the water molecules to increase the temperature to water's melting point, 0° C.

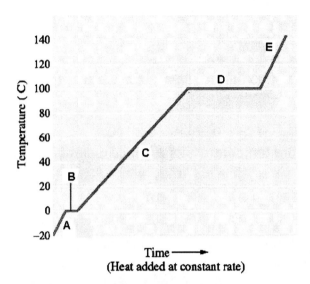

(Heat added at constant rate)

The slope of the line for this interval shows the increase in temperature and is related to the heat capacity of the substance.

During the interval marked B on the graph, energy is still being added to the water but the temperature remains the same, at 0° C or water's melting point temperature. The additional energy is being used to overcome the intermolecular forces holding the water molecules in their solid pattern. This energy is moving the particles apart, breaking or weakening the forces of attraction that keep the water molecules aligned. The solid water (ice) is being converted to liquid water; a phase change is occurring. The temperature will not increase until every solid particle has melted and the entire sample is liquid.

Temperature again increases during interval C on the graph. Energy is being absorbed by the liquid water molecules. Notice that the slope of the line during this interval is different than the slope of the line during interval A. This is due to differences in the heat capacity of ice and liquid water.

The flat line during interval D indicates that a phase change is occurring. The additional energy is being used to overcome the attractive forces holding the liquid water molecules together. The water molecules increase their kinetic energies and move farther apart, changing to water vapor. This occurs at the boiling point temperature, or 100° C. The temperature stays at the boiling point temperature until all water molecules are converted to water vapor. Once this conversion occurs, the temperature increases as energy is added, according to the heat capacity of the substance as a vapor.

Skill 13.7   Determining heat transfer using mass, specific heat, and temperature change

Heat is a measure of energy. If two objects that have different temperatures come into contact with each other, heat flows from the hotter object to the cooler.

Heat Capacity of an object is the amount of heat energy that it takes to raise the temperature of the object by one degree.

Heat capacity (C) per unit mass (m) is called **specific heat** (c):

$$c = \frac{C}{m} = \frac{Q/\Delta T}{m}$$

Specific heats for many materials have been calculated and can be found in tables. Another way to look at specific heat is to think of it as the heat required to change the temperature of one gram by one degree.

There are a number of ways that heat is measured. In each case, the measurement is dependent upon raising the temperature of a specific amount of water by a specific amount. These conversions of heat energy and work are called the **mechanical equivalent of heat**.

The **calorie** is the amount of energy that it takes to raise one gram of water one degree Celsius. The **kilocalorie** is the amount of energy that it takes to raise one kilogram of water by one degree Celsius. Food calories are kilocalories. In the International System of Units **(SI)**, the calorie is equal to 4.184 **joules**. A **British thermal unit (BTU)** = 252 calories = 1.054 kJ

A substance's **enthalpy of fusion** (ΔH_{fusion}) is the heat required to **change one mole from a solid to a liquid** by melting. This is also the heat released from the substance when it changes from a liquid to a solid by freezing.

A substance's **enthalpy of vaporization** ($\Delta H_{vaporization}$) is the heat required to **change one mole of a substance from a liquid to a gas** or the heat released by condensation.

A substance's **enthalpy of sublimation** ($\Delta H_{sublimation}$) is the heat required to change one mole directly from a solid to a gas by sublimation or the heat released by deposition.

0014 UNDERSTAND THE RELATIONSHIPS AMONG FORCE, MASS, AND THE MOTION OF OBJECTS

Skill 14.1 Comparing types and characteristics of forces (e.g., frictional, gravitational) and analyzing the effects of forces on objects

Dynamics is the study of the relationship between motion and the forces affecting motion. **Force** causes motion.

Mass and weight are not the same quantities. An object's **mass** gives it a reluctance to change its current state of motion. It is also the measure of an object's resistance to acceleration. The force that the earth's gravity exerts on an object with a specific mass is called the object's weight on earth. Weight is a force that is measured in Newtons. Weight (W) = mass times acceleration due to gravity (**W = mg**).

Newton's laws of motion:

Newton's first law of motion is also called the law of inertia. It states that an object at rest will remain at rest and an object in motion will remain in motion at a constant velocity unless acted upon by an external force.

Newton's second law of motion states that if a net force acts on an object, it will cause the acceleration of the object. The relationship between force and motion is Force equals mass times acceleration. **(F = ma).**

Newton's third law states that for every action there is an equal and opposite reaction. Therefore, if an object exerts a force on another object, that second object exerts an equal and opposite force on the first.

Momentum

Every object in motion has a property called momentum. The amount of momentum depends on the mass and velocity of the object: **M = mv.** As either the mass or the velocity increases, the momentum increases. The greater the momentum of an object, the more force it takes to stop it. An object at rest has zero momentum.

Friction

Surfaces that touch each other have a certain resistance to motion. This resistance is **friction.**

1. The materials that make up the surfaces will determine the magnitude of the frictional force.
2. The frictional force is independent of the area of contact between the two surfaces.
3. The direction of the frictional force is opposite to the direction of motion.
4. The frictional force is proportional to the normal force between the two surfaces in contact.

Static friction describes the force of friction of two surfaces that are in contact but do not have any motion relative to each other, such as a block sitting on an inclined plane. **Kinetic friction** describes the force of friction of two surfaces in contact with each other when there is relative motion between the surfaces.

When an object moves in a circular path, a force must be directed toward the center of the circle in order to keep the motion going. This constraining force is called **centripetal force**. Gravity is the centripetal force that keeps a satellite circling the earth.

Skill 14.2   Analyzing the relationship between the displacement, velocity, and acceleration of an object graphically, algebraically, and in written form

The science of describing the motion of bodies is known as **kinematics**. The motion of bodies is described using words, diagrams, numbers, graphs, and equations.

The following words are used to describe motion: vectors, scalars, distance, displacement, speed, velocity, and acceleration.

The two categories of mathematical quantities that are used to describe the motion of objects are scalars and vectors. **Scalars** are quantities that are fully described by magnitude alone. Examples of scalars are 5m and 20 degrees Celsius. **Vectors** are quantities that are fully described by magnitude *and* direction. Examples of vectors are 30m/sec, and 5 miles north.

Distance is a scalar quantity that refers to how much ground an object has covered while moving. **Displacement** is a vector quantity that refers to the object's change in position. It includes distance and direction.

Example:

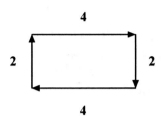

Jamie walked 2 miles north, 4 miles east, 2 miles south, and then 4 miles west. In terms of distance, she walked 12 miles. However, there is no displacement because the directions cancelled each other out, and she returned to her starting position.

Speed is a scalar quantity that refers to how fast an object is moving (ex. the car was traveling 60 mi/hr). **Velocity** is a vector quantity that refers to the rate at which an object changes its position. In other words, velocity is speed with direction (ex. the car was traveling 60 mi/hr east).

$$\text{Average speed} = \frac{\text{Distance traveled}}{\text{Time of travel}}$$

$$v = \frac{d}{t}$$

$$\text{Average velocity} = \frac{\triangle\text{position}}{\text{time}} = \frac{\text{displacement}}{\text{time}}$$

Instantaneous Speed - speed at any given instant in time.

Average Speed - average of all instantaneous speeds, found simply by a distance/time ratio.

Acceleration is a vector quantity defined as the rate at which an object changes its velocity.

$$a = \frac{\triangle velocity}{time} = \frac{v_f - v_i}{t}$$ where v_f represents the final velocity and v_i represents the initial velocity

Since acceleration is a vector quantity, it always has a direction associated with it. The direction of the acceleration vector depends on

- whether the object is speeding up or slowing down
- whether the object is moving in the positive or negative direction.

Adding Vectors

Quantities with magnitude and direction are **vector quantities** and can be represented, added, and subtracted graphically. If there are two vectors in the same direction, add them together. If there are two vectors in opposite directions, subtract the smaller from the larger.

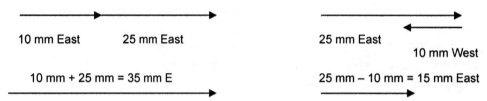

Vectors are not always this simple. However, it always helps to make a scaled drawing. If possible place the vectors head to tail. If they cannot simply be added or subtracted, you may need to use geometrical calculations based on the properties of right triangles:

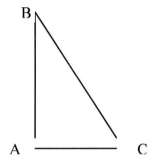

Angle A is the right angle. BC is the hypotenuse. To get the length of AB if you know angle B, use a cosine or to get the length of AB if you know angle C, use a sine. Sine or sin C = $^{\text{opposite side}}/_{\text{hypotenuse}}$ = $^{AB}/_{BC}$

Cosine or cos B = $^{\text{adjacent side}}/_{\text{hypotenuse}}$ = $^{AB}/_{BC}$

Tangent B or tan B = $^{\text{opposite side}}/_{\text{adjacent side}}$ = $^{AC}/_{AB}$

*Displacement, velocity, and acceleration are vector quantities.

Skill 14.3   Applying Newton's laws of motion to everyday situations

Push and pull –Pushing a volleyball or pulling a bowstring applies muscular force when the muscles expand and contract. Elastic force is when any object returns to its original shape (for example, when a bow is released).

Rubbing – Friction opposes the motion of one surface past another. Friction is common when slowing down a car or sledding down a hill.

Pull of gravity – is a force of attraction between two objects. Gravity questions can be raised not only on earth but also between planets and even black hole discussions.

Forces on objects at rest – The formula F= m/a is shorthand for force equals mass over acceleration. An object will not move unless the force is strong enough to move the mass. Also, there can be opposing forces holding the object in place. For instance, a boat may want to be forced by the currents to drift away but an equal and opposite force is a rope holding it to a dock.

Forces on a moving object - Overcoming inertia is the tendency of any object to oppose a change in motion. An object at rest tends to stay at rest. An object that is moving tends to keep moving.

Inertia and circular motion – The centripetal force is provided by the high banking of the curved road and by friction between the wheels and the road. This inward force that keeps an object moving in a circle is called centripetal force.

Skill 14.4   Solving problems related to motion, force, and momentum

Using these laws of motion and algebra, we can solve for force, mass or acceleration, provided that we have two of the three variables. The force of gravity on earth is always equal to the weight of the object as found by the equation:

Fgrav = m * g

where g = 9.8 m/s^2 (on Earth)
and m = mass (in kg)

Momentum is the amount of moving mass. Its SI unit is kg m/s because it is calculated as the mass multiplied by the velocity of an object. Momentum is the mathematical restatement of Newton's laws- it is the tendency of an object to continue to move in its direction of travel, unless acted on by a net external force. Momentum is 'conserved' (the total momentum of a closed system cannot be changed.

Example: Which has more momentum, a 3600-kilogram truck moving at 8 kilometers per hour or an 1800-kilogram car moving at 16 kilometers per hour? If they both move at the same speed, which one will have the greater momentum?

For the truck: M = mv = 3600 kg x 8 km/hr = 28,800 kg-km/hr
For the car: M = mv = 1800 kg x 16 km/hr = 28,800 kg-km/hr
If the car moves at 8 km: M = mv = 1800 kg x 8 km/hr = 14,400 kg-km/hr

Under the given conditions of the truck moving at 8 km/hr and the car moving at 16 km/hr, the momentum of both is identical. However, if both move at the same speed, the truck has greater momentum since it has a greater mass.

$$\frac{\overset{4}{3600}}{\underset{28\cancel{8}.00}{8}} \ kg-km$$

If you are going to drive a car, you start from being parked. The car at rest is at a speed of zero. You apply pressure to the accelerator or gas pedal and the car starts moving. In less than a minute, it may reach a speed of 50 km per hour. The equation for **acceleration** is:

$$a = {}^{v2 - v1}/_t \text{ where v1 is the beginning velocity of 0}$$

Likewise, when you come to a red light, you apply the brake and use the force of braking to decelerate to zero. To calculate the **deceleration**, you would use the same equation as for acceleration, but v2 would be zero instead of v1 being zero. Acceleration and deceleration are rates at which speed changes.

Falling objects accelerate at a rate of 9.8 meters per second squared (9.8 m/s^2). That means that the speed of the falling object increases 9.8 meters per second for each second it falls. If it starts at rest, its speed the first second is 9.8 meters per second, its speed for the next second is 19.6 meters per second (2 x 9.8), its speed for the third second is 29.4 meters per second (3 x 9.8), and so forth.

Example: If a ball is released from rest and drops for 3 sec., what is its (a) initial velocity, (b) final velocity, and (c) average velocity?
If it is released from rest, its initial velocity is 0.
Its final velocity is 3 x 9.8 m/s^2 or 29.4 m/s^2
Its average velocity is $v_{av} = (v + v_0) / 2 = (29.4 \text{ m/s}^2 + 0)/2 = 14.7 \text{ m/s}^2$

Example: If a ball is dropped from a height of 54 m, how long will it take to hit the ground and what will its velocity be just before it strikes the ground?
Given: $a = 9.8 \text{ m/s}^2$, d = 54 m, and $v_0 = 0$

$$d = v_0 t + \tfrac{1}{2} a t^2$$

$O = 5$

$54 \text{ m} = 0 + \tfrac{1}{2} (9.8 \text{ m/s}^2)(t^2)$
$t^2 = 11.0 \text{ s}^2$

$O \, km/hr = 54' m \bullet 9.8 m s^2$

$t = 3.32 \text{ seconds}$

$$v^2 = v_0^2 + 2ad$$

$v^2 = 0 + (2)(9.8 \text{ m/s}^2)(54 \text{ m})$
$v^2 = 1,060 \text{ m}^2/\text{s}^2$
$v = 32.6 \text{ m/s}$

In a vacuum there is no air, so there is no air resistance. However, in a normal situation you will see something like a feather drop more slowly due to **air resistance**, the upward force of air against the object.

When an arrow is shot horizontally from a bow, it will move in a horizontal motion but it will also be pulled down by gravity. Therefore, the archer will have to allow for the downward velocity and aim above the bulls-eye on the target.

Example: If a boat is to start on one side of a river flowing 3 km/hr and end up directly across river from where it started, it will have to be pointed upstream at some angle. The boat will travel 6 km/hr. What will its velocity need to be relative to the earth?

Tan θ = opp/adj = 3/6 = .5
θ = 26.5° upstream
Sin θ = opp/hyp
sin 26.5° = 3 km/hr / hyp
0.446 = 3/hyp
Hyp = 3/.446 = 6.7 km/hr

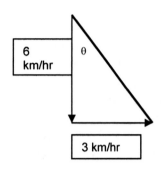

Skill 14.5   Applying knowledge of the concepts of work, power, efficiency, and mechanical advantage

Work and energy:

Work is done on an object when an applied force moves through a distance.

Work is measured in newton-meters, also called joules:

$$W = Fd$$

Whenever work is done upon an object by an external force, there will be a change in the total mechanical energy of the object. If only internal forces are doing work (no work done by external forces), there is no change in total mechanical energy, the total mechanical energy is "conserved." The quantitative relationship between work and mechanical energy is expressed by the following equation:

$$TME_i + W_{ext} = TME_f$$

The equation states that the initial amount of total mechanical energy (TME_i) plus the work done by external forces (W_{ext}) is equal to the final amount of total mechanical energy (TME_f).

Power

Power is the rate at which work is done. It is the work/time ratio. The following equation is the formula used to compute power:

$$Power = \frac{Work}{Time}$$

The standard metric unit of power is the Watt. A unit of power is equivalent to a unit of work divided by a unit of time. Therefore, a Watt is equivalent to a Joule/second. When force and velocity are constant, the instantaneous power is equal to the average power.

According to the **Law of Conservation of Energy,** energy cannot be created or destroyed; it may be transformed from one form into another, but the total amount of energy never changes. Energy can come from a variety of sources: fossil fuels, solar energy, hydroelectric power, biomass, wind, geothermal, and nuclear fission and fusion. Mass in each of these is transformed to energy through the equation $E = mc^2$ where E is energy, m is mass, and c is the speed of light. Solar energy in the form of radiation is absorbed by plants to make chemical energy through photosynthesis. This energy is stored by the plant to be used by an animal eating the plant or to be converted into light and heat many years later after the plant has become part of a peat or coal bed.

Energy can be divided into two types. **Kinetic energy**, the energy of motion, is seen in a bowling ball knocking down the bowling pins. Turning on a flashlight demonstrates the stored energy, or **potential energy**, of the batteries. Potential energy is also known as positional energy. For example, a rock at the top of a hill has potential energy but a rock in motion, rolling down the hill, has kinetic motion.

To find the kinetic energy of a body, consider the work which must be done on the body in order to give it its speed. When the body is stopped, it gives up this amount of energy. Therefore, **KE = ½ mv²**.

Example: If a 1,000 kg automobile is moving with a speed of 20 m/s, what is the kinetic energy in joules?

$$KE = \frac{1}{2} mv^2$$
$$KE = \frac{1}{2} (1,000 \text{ kg})(20 \text{ m/s})^2$$
$$KE = 2 \times 10^5 \text{ J}$$

The measure of the potential energy which a body has because of its elevated position is the work done against gravity in lifting the body from some level chosen as the zero for potential energy. The upward force required is equal to the product of the weight of the body (w) and the work done in lifting the body through a height (h) such that **PE = wh = mgh**.

Example: What is the potential energy of a 50-kg hammer of a pile driver when it is raised 4 m?

$$PE = wh = mgh$$
$$PE = 50 \text{ kg} \times 9.8 \text{ m/s}^2 \times 4 \text{ m}$$
$$PE = 1,960 \text{ J (Using significant figures PE} = 2.0 \times 10^3 \text{ J)}$$

Another important example of potential energy is the energy in a compressed spring. In general, the force that is applied to the spring is proportional to the extension of the spring: **F = ky** where k is a spring constant and y is the extension. To produce a displacement the applied force is zero at first and increases linearly to ky. The work done by the force is the product of the displacement (extension) and the average force (ky/2): PE = ½ ky^2.

In many cases potential energy is transformed directly into kinetic energy. One example is water going over a dam. Another is a pendulum swinging. When the pendulum is at its highest point on the right or the left, that is its greatest potential energy and when it is vertical (its equilibrium position) it is at its greatest kinetic energy. Therefore, it goes from PE to KE to PE to KE and so forth.

Example: A pendulum bob is pulled to one side until its center of gravity has been raised 10 cm above its equilibrium position. Find the speed of the bob as it swings through the equilibrium position.

$$PE \text{ at top} = KE \text{ at bottom}$$
$$mhg = \tfrac{1}{2} mv^2 \text{ or } 2hg = v^2$$
$$(2)(0.1 \text{ m})(9.8 \text{ m/s}^2) = v^2$$
$$1.96 \text{ m}^2/\text{s}^2 = v^2$$
$$1.4 \text{ m/s} = v$$

(handwritten margin notes:) 10 ⋆   10   .1 m (9.8)

Mechanical Advantage

There are two types of mechanical advantage, ideal and actual. Ideal mechanical advantage is the mechanical advantage of an ideal machine. Because such a machine does not really exist, we use physics principals to "theoretically" solve such equations. The ideal mechanical advantage (IMA) is found using the following formula:

$$IMA = D_E / D_R$$

The effort distance divided by the resistance distance gives us the IMA.

Actual mechanical advantage it the mechanical advantage of a real machine and takes into consideration factors such as energy lost to friction. Actual mechanical advantage (AMA) is calculated using the following formula:

$$AMA = R / E_{actual}$$

Dividing the resistance force by the actual effort force we can determine the actual mechanical advantage of a machine.

Efficiency is the relationship between energy input and energy output. Efficiency is expressed as a percentage. The more efficient a system is, the less energy that is lost within that system. The percentage efficiency of any machine can be calculated as long as you know how much energy has to be put into the machine and how much useful energy comes out. Use the following equation:

% efficiency = useful energy produced x 100 / total energy used

Skill 14.6   Identifying types and characteristics of simple machines

Simple machines include the following:

1. Inclined plane
2. Lever
3. Wheel and axle
4. Pulley
5. Screw
6. Wedge

Levers have two parts – a resistance arm and an effort arm. The effort arm is the distance from the fulcrum to the effort force. The resistance arm is the distance from the fulcrum to the resistance force. The resistance force is the weight of the object the lever will move.

There are three classes of levers. They are based on the position of the fulcrum, resistance and effort. In a first class lever, the fulcrum is between the effort and resistance. Examples include shoveling snow, cutting with scissors, and pulling a nail with the claw of a hammer. A second class lever has the resistance between the effort and the fulcrum. Examples of this are a wheelbarrow, a nutcracker, or a bottle opener to open glass bottles. The third class lever has the effort between the fulcrum and resistance. Raking leaves, hitting a baseball with a bat and using a fishing rod are examples.

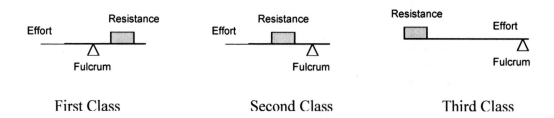

First Class   Second Class   Third Class

A pulley makes work easier by changing the direction of the force. The effort force is applied in one direction and the resistance moves in the opposite direction with a fixed pulley. Two or more pulleys can be used together to decrease the force needed to move an object. The mechanical advantage of a set of pulleys is about equal to the number of supporting ropes or strands.

GENERAL SCIENCE   152

The wheel and axle acts like a first class lever. The effort arm is the radius of the wheel. The resistance arm is the radius of the axle. The fulcrum is the center of the axle. The mechanical advantage of the wheel and axle is equal to the radius of the wheel divided by the radius of the axle.

An inclined plane (such as a ramp) is a slanted surface used to raise or lower objects. Less force is needed to lift an object using an inclined plane. However the object is moved through a longer distance than if it were moved straight up. The mechanical advantage of an inclined plane is its length divided by its height. The longer the inclined plane, the less force is needed to move the object.

A wedge is two inclined planes. A knife blade and an ax blade are wedges. The wedge moves in doing work, but a simple inclined plane does not move. A screw is also a form of inclined plane. It is like a spiral staircase with the steps wound around the center.

Simple machines make work easier. **Compound machines** are two or more simple machines working together. A wheelbarrow is an example of a compound machine. It uses a lever and a wheel and axle. Machines of all types ease workload by changing the size or direction of an applied force. The amount of effort saved when using simple or complex machines is called mechanical advantage or MA.

Efficiency of a machine is increased by decreasing friction. Therefore, sanding rough edges or greasing bearings on machines decreases friction and conserves energy.

0015 UNDERSTAND THE PROPERTIES OF WAVES, SOUND, AND LIGHT

Skill 15.1   Comparing and contrasting characteristics of longitudinal waves and transverse waves

A wave is a rhythmic disturbance which travels through space or matter. By tying a rope to a post, you can move your end of the rope up and down to create wave action. Just as waves appear to move across the rope but the particles of the rope do not move horizontally to create the waves, waves of other types (water, for example) are made of particles that oscillate up and down rather moving across a space.

Waves are classified in terms of how the motion of the individual particles of the medium is related to the movement of the wave itself. In waves produced by the rope, the particles move up and down at right angles to the direction in which the wave itself moves. **Transverse waves** are characterized by the particle motion being perpendicular to the wave motion; **longitudinal or compressional waves** are characterized by the particle motion being parallel to the wave motion. Sound waves are a good example of longitudinal waves.

Transverse Wave

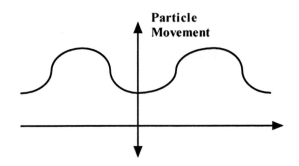

Direction of Energy Transport

Longitudinal Wave

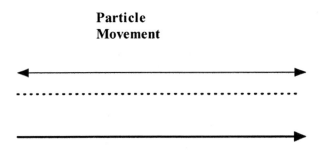

Direction of Energy Transport

The parts of a wave are:

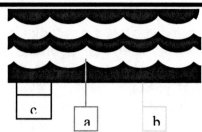

Crest (a) – top of the hill

Trough (b) – bottom of the valley

Wavelength (c) – the total distance from the center of one crest through one trough to the center of the next crest

Amplitude – the distance a wave rises or falls from its usual rest position

Frequency – the number of waves that pass a given point in one second expressed in hertz (Hz) which means 'per second'

As the length of a wave increases, its frequency decreases. Long waves have a low frequency while short waves have a high frequency.

The **speed of a wave** depends on the wavelength and frequency of the wave.
Speed (V) = wavelength (λ) x frequency (f)

Example: A tuning fork has a frequency of 256 Hz. The wavelength of the sound wave produced by the fork is about 1.3 meters. What is the speed of the wave?

$$V = \lambda f$$
$$V = 1.3 \text{ m} \times 256 \text{ Hz}$$
$$V = 332.8 \text{ m/s}$$

Skill 15.2 **Analyzing how the behavior of waves is affected by the medium (e.g., air, water, solids) through which the waves are passing**

Water waves in a tank can be used as models to explain the behavior of sound, light and other energy waves. Waves that strike an object are called **incident waves**. The waves that bounce off are called **reflected waves**. The angle between the wave and a line drawn normal to the wall is the angle of incidence. A line drawn normal to the wall means the line is perpendicular to the wall at the point where the wave strikes the wall. The angle between the reflected wave and the normal is the angle of reflection. The angle of incidence equals the angle of reflection.

Refraction is the bending of waves. As water waves pass into shallow water, their wavelength becomes shorter. The speed of the waves decreases. The incident wave is bent because the whole wave is not traveling at the same speed.

Sound waves need a medium in which to spread; light waves do not. The speed of any wave depends upon the elastic and inertial properties of the medium through which it travels. The density of a medium is an example of an **inertial property**. Sound usually travels faster in denser material. A sound wave will travel nearly three times as fast through helium as through air. On the other hand, the speed of light is slower in denser materials. The speed of light is slower in glass than in air. (The standard for the speed of light, c, is actually the speed of light in a vacuum, such as empty space.)

Elastic properties are properties related to the tendency of a medium to maintain its shape when acted upon by force or stress. Sound waves travel faster in solids than they do in liquids, and faster in liquids than they do in gases. The inertial factor would seem to indicate otherwise. However, the elastic factor has a greater influence on the speed of the wave.

When a wave strikes an object, some of the wave energy is reflected off the object, some of the energy goes into and is absorbed by the object, and some of the energy goes through the object. For example, sound waves can penetrate walls. However, sound waves from the air cannot penetrate water, and sound waves from water cannot penetrate the air. Light passes through some materials such as glass but not many other materials.

Skill 15.3   Analyzing the phenomena of reflection, refraction, interference, diffraction, polarization, dispersion, and absorption

Shadows illustrate one of the basic properties of light. Light travels in a straight line. If you put your hand between a light source and a wall, you will interrupt the light and produce a shadow.

When light hits a surface, it is **reflected.** The angle of the incoming light (angle of incidence) is the same as the angle of the reflected light (angle of reflection). It is this reflected light that allows you to see objects. You see the objects when the reflected light reaches your eyes.

Different surfaces reflect light differently. Rough surfaces scatter light in many different directions. A smooth surface reflects the light in one direction. If it is smooth and shiny (like a mirror) you see your image in the surface.

When light enters a different medium, it bends. This bending, or change of speed, is called **refraction**.

Wave **interference** occurs when two waves meet while traveling along the same medium. The medium takes on a shape resulting from the net effect of the individual waves upon the particles of the medium. There are two types of interference: constructive and destructive.

Constructive interference occurs when two crests or two troughs of the same shape meet. The medium will take on the shape of a crest or a trough with twice the amplitude of the two interfering crests or troughs. If a trough and a crest of the same shape meet, the two pulses will cancel each other out, and the medium will assume the equilibrium position. This is called **destructive** interference.

Destructive interference in sound waves will reduce the loudness of the sound. This is a disadvantage in rooms, such as auditoriums, where sound needs to be at its optimum. However, it can be used as an advantage in noise reduction systems. When two sound waves differing slightly in frequency are superimposed, beats are created by the alternation of constructive and destructive interference. The frequency of the beats is equal to the difference between the frequencies of the interfering sound waves.

Wave interference occurs with light waves in much the same manner that it does with sound waves. If two light waves of the same color, frequency, and amplitude are combined, the interference shows up as fringes of alternating light and dark bands. In order for this to happen, the light waves must come from the same source.

Light can be **diffracted**, or bent, around the edges of an object. Diffraction occurs when light goes through a narrow slit. As light passes through it, the light bends slightly around the edges of the slit. You can demonstrate this by pressing your thumb and forefinger together, making a very thin slit between them. Hold them about 8 cm from your eye and look at a distant source of light. The pattern you observe is caused by the diffraction of light.

A great example **absorption** is apparent when discussing light. Transparent materials allow one or more of the frequencies of visible light to be transmitted through them; whatever color(s) is/are not transmitted by such objects, are typically absorbed by them. The appearance of a transparent object is dependent upon what color(s) of light is/are incident upon the object and what color(s) of light is/are transmitted through the object.

Light can be **polarized** because the waves are transverse. The distinguishing characteristic of transverse waves is that they are perpendicular to the direction of the motion of the wave. Polarized light has vibrations confined to a single plane that is perpendicular to the direction of motion. Light is able to be polarized by passing it through special filters that block all vibrations except those in a single plane. By blocking out all but one place of vibration, polarized sunglasses cut down on glare. Other types of electromagnetic radiation can also be polarized.

Skill 15.4   Demonstrating knowledge of characteristics and uses of electromagnetic radiation

The electromagnetic spectrum is measured in frequency (f) in hertz and wavelength (λ) in meters. The frequency times the wavelength of every electromagnetic wave equals the speed of light (3.0×10^9 meters/second).

Roughly, the range of wavelengths of the electromagnetic spectrum is:

	$\underline{f}$	λ
Radio waves	$10^5 - 10^{-1}$ hertz	$10^3 - 10^9$ meters
Microwaves	$10^{-1} - 10^{-3}$ hertz	$10^9 - 10^{11}$ meters
Infrared radiation	$10^{-3} - 10^{-6}$ hertz	$10^{11.2} - 10^{14.3}$ meters
Visible light	$10^{-6.2} - 10^{-6.9}$ hertz	$10^{14.3} - 10^{15}$ meters
Ultraviolet radiation	$10^{-7} - 10^{-9}$ hertz	$10^{15} - 10^{17.2}$ meters
X-Rays	$10^{-9} - 10^{-11}$ hertz	$10^{17.2} - 10^{19}$ meters
Gamma Rays	$10^{-11} - 10^{-15}$ hertz	$10^{19} - 10^{23.25}$ meters

Radio waves are used for transmitting data. Common examples are television, cell phones, and wireless computer networks. Microwaves are used to heat food and deliver Wi-Fi service. Infrared waves are utilized in night vision goggles. Visible light we are all familiar with as the human eye is most sensitive to this wavelength range. UV light causes sunburns and would be even more harmful if most of it were not captured in the Earth's ozone layer. X-rays aid us in the medical field and gamma rays are most useful in the field of astronomy.

Light can travel through thin fibers of glass or plastic without escaping the sides. Light on the inside of these fibers is reflected so that it stays inside the fiber until it reaches the other end. Such **fiber optics** are being used to carry telephone messages. Sound waves are converted to electric signals, which are coded into a series of light pulses which then proceed through the optical fiber until they reach the other end. At that time, they are converted back into sound.

The image that you see in a bathroom mirror is a **virtual image** because it only seems to be where it is. However, a curved mirror can produce a real image. A real image is produced when light passes through the point where the image appears. A **real image** can be projected onto a screen.

Cameras use a convex lens to produce an image on the film. A **convex lens** is thicker in the middle than at the edges. The image size depends upon the focal length (distance from the focus to the lens); the longer the focal length, the larger the image. A **converging lens** produces a real image whenever the object is far enough from the lens so that the rays of light from the object can hit the lens and be focused into a real image on the other side of the lens.

Eyeglasses can help correct deficiencies of sight by changing where the image seen is focused on the retina of the eye. If a person is nearsighted, the lens of his eye focuses images in front of the retina. In this case, the corrective lens placed in the eyeglasses will be concave so that the image will reach the retina. In the case of farsightedness, the lens of the eye focuses the image behind the retina. The correction will call for a convex lens to be fitted into the glass frames so that the image is brought forward into sharper focus.

Skill 15.5 Demonstrating knowledge of the properties of sound and light in everyday phenomena (e.g., echoes, Doppler effect, magnification, rainbows)

Sound waves are produced by a vibrating body. The vibrating object moves forward and compresses the air in front of it, then reverses direction so that the pressure on the air is lessened and expansion of the air molecules occurs. One compression and expansion creates one longitudinal wave. Sound can be transmitted through any gas, liquid, or solid. However, it cannot be transmitted through a vacuum, because there are no particles present to vibrate and bump into their adjacent particles to transmit the wave.

The vibrating air molecules move back and forth parallel to the direction of motion of the wave as they pass the energy from adjacent air molecules (closer to the source) to air molecules farther away from the source.

The **amplitude** of a sound wave determines its loudness. Loud sound waves have large amplitudes. The larger the sound wave, the more energy is needed to create the wave.

When a piano tuner tunes a piano, he only uses one tuning fork, even though there are many strings on the piano. He adjusts to first string to be the same as that of the tuning fork. Then he listens to the beats that occur when both the tuned and untuned strings are struck. He adjusts the untuned string until he can hear the correct number of beats per second. This process of striking the untuned and tuned strings together and timing the beats is repeated until all the piano strings are tuned.

Change in experienced frequency due to relative motion of the source of the sound is called the **Doppler Effect.** When a siren approaches, the pitch is high. When it passes, the pitch drops. As a moving sound source approaches a listener, the sound waves are closer together, causing an increase in frequency in the sound that is heard. As the source passes the listener, the waves spread out and the sound experienced by the listener is lower.

The most common type of **microscope** is the optical microscope. This is an instrument containing one or more lenses that produce an enlarged (**magnified**) image of an object placed in the focal plane of the lens(es). Microscopes can largely be separated into two classes: optical theory microscopes and scanning microscopes. Optical theory microscopes are microscopes which function through lenses to magnify the image generated by the passage of a wave through the sample, this is what one thinks of in the high school classroom. The waves used are either electromagnetic in optical microscopes or electron beams in electron microscopes. Common types are the Compound Light, Stereo, and the electron microscope. Optical microscopes use refractive lenses, typically of glass and occasionally of plastic, to focus light into the eye or another light detector. Typical magnification of a light microscope is up to 1500x. Electron microscopes, which use beams of electrons instead of light, are designed for very high magnification and resolution. The most common of these would be the scanning electron microscopes used in spectroscopy studies.

A **rainbow** is an optical phenomenon. The rainbow's appearance is caused by dispersion of sunlight as it is refracted by raindrops. Hence, rainbows are commonly seen after a rainfall, or near fountains and waterfalls. A rainbow does not actually exist at a specific location in the sky, but rather is an optical phenomenon whose apparent position depends on the observer's location. All raindrops refract and reflect the sunlight in the same way, but only the light from some raindrops will reach the observer's eye. These raindrops create the perceived rainbow (as experienced by that observer).

An **echo** is a wave that has been reflected by a medium, and returns to your ear. The delay between its reflection and your perception of its return is equal to the distance divided by the speed of sound.

Skill 15.6 Demonstrating knowledge of the relationship between the properties of waves and how they are perceived by humans (e.g., color, pitch)

When we refer to light, we are usually talking about a type of electromagnetic wave that stimulates the retina of the eye, or visible light. Each individual wavelength within the spectrum of visible light represents a particular **color**. When a particular wavelength strikes the retina, we perceive that color. Visible light is sometimes referred to as ROYGBIV (red, orange, yellow, green, blue, indigo, violet). The visible light spectrum ranges from red (the longest wavelength) to violet (the shortest wavelength) with a range of wavelengths in between. If all the wavelengths strike your eye at the same time, you will see white. Conversely, when no wavelengths strike your eye, you perceive black.

The **pitch** of a sound depends on the **frequency** that the ear receives. High-pitched sound waves have high frequencies. High notes are produced by an object that is vibrating at a greater number of times per second than one that produces a low note.

The **intensity** of a sound is the amount of energy that crosses a unit of area in a given unit of time. The loudness of the sound is subjective and depends upon the effect on the human ear. Two tones of the same intensity but different pitches may appear to have different loudness. The intensity level of sound is measured in decibels. Normal conversation is about 60 decibels. A power saw is about 110 decibels.

The **amplitude** of a sound wave determines its loudness. Loud sound waves have large amplitudes. The larger the sound wave, the more energy is need to create the wave.

Pleasant sounds have a regular wave pattern that is repeated over and over. Sounds that do not happen with regularity are unpleasant and are called **noise**.

TEACHER CERTIFICATION STUDY GUIDE

0016 UNDERSTAND ELECTRICITY AND MAGNETISM

Skill 16.1   Identifying the characteristics of static electricity and explaining how it is generated

Electrostatics is the study of stationary electric charges. A plastic rod that is rubbed with fur or a glass rod that is rubbed with silk will become electrically charged and will attract small pieces of paper. The charge on the plastic rod rubbed with fur is negative and the charge on glass rod rubbed with silk is positive. The plastic rod gathers electrons while the glass rod loses electrons.

Electrically charged objects share these characteristics:

1. Like charges repel one another.
2. Opposite charges attract each other.
3. Charge is conserved.

A neutral object has no net change. If the plastic rod and fur are initially neutral, when the rod becomes charged by the fur, a negative charge is transferred from the fur to the rod. The net negative charge on the rod is equal to the net positive charge on the fur.

Materials through which electric charges can easily flow are called **conductors**. Metals which are good conductors include silicon and boron. On the other hand, an **insulator** is a material through which electric charges do not move easily, if at all. An example of an insulator would be non-metal elements of the periodic table. A simple device used to indicate the existence of a positive or negative charge is called an **electroscope**. An electroscope is made up of a conducting knob and attached to it are very lightweight conducting leaves usually made of foil (gold or aluminum). When a charged object touches the knob, the leaves push away from each other because like charges repel. It is not possible to tell whether if the charge is positive or negative.

Charging by induction:

Touch the knob with a finger while a charged rod is nearby. The electrons will be repulsed and flow out of the electroscope through the hand. If the hand is removed while the charged rod remains close, the electroscope will retain the charge.

When an object is rubbed with a charged rod, the object will take on the same charge as the rod. However, charging by induction gives the object the opposite charge as that of the charged rod.

GENERAL SCIENCE   162

Grounding charge:

Charge can be removed from an object by connecting it to the earth through a conductor. The removal of static electricity by conduction is called **grounding**.

Skill 16.2   Applying knowledge of the flow of electrons in circuits, including the relationships between potential difference, resistance, and current

An **electric circuit** is a path along which electrons flow. A simple circuit can be created with a dry cell, wire, and a device such as a bell or a light bulb. When all are connected, the electrons flow from the negative terminal, through the wire to the device and back to the positive terminal of the dry cell. If there are no breaks in the circuit, the device will work. The circuit is closed. Any break in the flow will create an open circuit and cause the device to shut off.

The device (bell, bulb) is an example of a **load**. A load is any device that uses energy. Suppose that you also add a buzzer so that the bell rings when you press the buzzer button. The buzzer is acting as a **switch**. A switch is a device that opens or closes a circuit. Pressing the buzzer makes the connection complete and the bell rings. When the buzzer is not engaged, the circuit is open and the bell is silent.

When an electron goes through a load, it does work and, therefore, loses some of its energy. The measure of how much energy is lost is called the **potential difference**. The potential difference between two points is the work needed to move a charge from one point to another.

Potential difference is measured in a unit called the volt. **Voltage** is potential difference. The higher the voltage, the more energy the electrons have. This energy is measured by a voltmeter. To use a voltmeter, place it in a circuit <u>parallel</u> with the load you are measuring.

Current is the number of electrons per second that flow past a given point in a circuit. Current is measured with an ammeter. To use an ammeter, put it in <u>series</u> with the load you are measuring. The magnitude of the current (I) is the charge per unit of time that passes any cross section of wire:
I = Q / t. Current flows from a point at higher potential to a point at lower potential as though the current represented a movement of positive charge even though it is the movement of negatively charged electrons

As electrons flow through a wire, they lose potential energy. Some is changed into heat energy because of resistance. **Resistance** is the ability of the material to oppose the flow of electrons through it. All substances have some resistance, even if they are good conductors such as copper. This resistance is measured in units called **ohms**. A thin wire will have more resistance than a thick one because it will have less room for electrons to travel. In a thicker wire, there will be more possible paths for the electrons to flow. Resistance also depends upon the length of the wire. The longer the wire, the more resistance it will have. Potential difference, resistance, and current form a relationship know as **Ohm's Law**. Current **(I)** is measured in amperes and is equal to potential difference **(V)** divided by resistance **(R)**.

$$I = V / R$$

The power in a circuit is the voltage of the source multiplied by the current produced by the source. Voltage is joules per coulomb. The result of the multiplication is joules per second which is watts.

Power (watts) = voltage (J/C) x current (C/s)
$$P = V \, I$$

Therefore, $P = I^2 R$ or $P = V^2 / R$
As a result: $W = I^2 R \, t$

If you have a wire with resistance of 5 ohms and a potential difference of 75 volts, you can calculate the current by

I = 75 volts / 5 ohms
I = 15 amperes

A current of 10 or more amperes will cause a wire to get hot. 22 amperes is about the maximum for a house circuit. Anything above 25 amperes can start a fire.

Skill 16.3   Comparing and contrasting series and parallel circuits and how they transfer energy

A **series circuit** is one where the electrons have only one path along which they can move. When one load in a series circuit goes out, the circuit is open. An example of this is a set of Christmas tree lights that is missing a bulb. None of the bulbs will work.

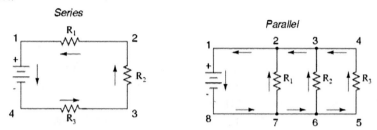

A **parallel circuit** is one where the electrons have more than one path to move along. If a load goes out in a parallel circuit, the other load will still work because the electrons can still find a way to continue moving along the path.

Skill 16.4   Recognizing the characteristics and uses of magnetic domains, magnets, and magnetic fields

Magnets have a north pole and a south pole. Like poles repel and opposing poles attract. A **magnetic field** is the space around a magnet where its force will affect objects. The closer you are to a magnet, the stronger the force. As you move away, the force becomes weaker.

Some materials act as magnets and some do not. This is because magnetism is a result of electrons in motion. The most important motion in this case is the spinning of the individual electrons. Electrons spin in pairs in opposite directions in most atoms. Each spinning electron has the magnetic field that it creates canceled out by the electron that is spinning in the opposite direction.

In an atom of iron, there are four unpaired electrons. The magnetic fields of these are not canceled out. Their fields add up to make a tiny magnet. There fields exert forces on each other setting up small areas in the iron called **magnetic domains** where atomic magnetic fields line up in the same direction.

You can make a magnet out of an iron nail by stroking the nail in the same direction repeatedly with a magnet. This causes poles in the atomic magnets in the nail to be attracted to the magnet. The tiny magnetic fields in the nail line up in the direction of the magnet. The magnet causes the domains pointing in its direction to grow in the nail. Eventually, one large domain results and the nail becomes a magnet.

A bar magnet has a north pole and a south pole. If you break the magnet in half, each piece will have a north and south pole.

The earth has a magnetic field. In a compass, a tiny, lightweight magnet is suspended and will line its south pole up with the geophysical magnetic North Pole of the earth.

Skill 16.5   Demonstrating knowledge of the relationship between electricity and magnetism and applications of electromagnetism and electromagnetic induction (e.g., motors, generators, transformers)

A magnet can be made out of a coil of wire by connecting the ends of the coil to a battery. When the current goes through the wire, the wire acts in the same way that a magnet does, it is called an **electromagnet**. The poles of the electromagnet will depend upon which way the electric current runs. An electromagnet can be made more powerful in three ways:

1. Make more coils.
2. Put an iron core (nail) inside the coils.
3. Use more battery power.

An **electric meter**, such as the one found on the side of a house, contains an aluminum disk that sits directly in a magnetic field created by electricity flowing through a conductor. The more the electricity flows (current), the stronger the magnetic field is. The stronger the magnetic field, the faster the disk turns. The disk is connected to a series of gears that turn a dial. Meter readers record the number from that dial.

In a **motor**, electricity is used to create magnetic fields that oppose each other and cause the rotor to move. The wiring loops attached to the rotating shaft have a magnetic field opposing the magnetic field caused by the wiring in the housing of the motor that cannot move. The repelling action of the opposing magnetic fields turns the rotor.

Telegraphs use electromagnets to work. When a telegraph key is pushed, current flows through a circuit, turning on an electromagnet which attracts an iron bar. The iron bar hits a sounding board, which responds with a click. Release the key and the electromagnet turns off. Messages can be sent around the world in this way.

Scrap metal can be removed from waste materials by the use of a large electromagnet that is suspended from a crane. When the electromagnet is turned on, the metal in the pile of waste will be attracted to it. All other materials will stay on the ground.

Air conditioners, vacuum cleaners, and washing machines use electric motors. An electric motor uses an electromagnet to change electric energy into mechanical energy.

A **generator** is a device that turns rotary mechanical energy into electrical energy. The process is based on the relationship between magnetism and electricity. As a wire or any other conductor moves across a magnetic field, an electric current occurs in the wire. The large generators used by the electric companies have a stationary conductor. A magnet attached to the end of a rotating shaft is positioned inside a stationary, conducting ring that is wrapped with a long, continuous piece of wire. When the magnet rotates, it induces a small electric current in each section of wire as it passes. Each section of wire is a small, separate electric conductor. All the small currents of these individual sections add up to one large current, which is what is used for electric power.

A **transformer** is an electrical device that changes electricity of one voltage into another voltage, usually from high to low voltage. You can see transformers at the top of utility poles. A transformer uses two properties of electricity: first, magnetism surrounds an electric circuit and second, voltage is made when a magnetic field moves or changes strength. Voltage is a measure of the strength or amount of electrons flowing through a wire. If another wire is close to an electric current changing strength, the electric current will also flow into that other wire as the magnetism changes. A transformer takes in electricity at a higher voltage and lets it run through many coils wound around an iron core. The magnetism in the core alternates because the current is alternating. An output wire with fewer coils is also around the core. The changing magnetism makes a current in the output wire. Having fewer coils means less voltage, so the voltage is reduced.

Skill 16.6 Identifying the processes involved in the transformation of mechanical energy into electrical energy and the transmission of electrical energy

Almost any form of energy can be transformed into another form. Indeed, the generation of electricity for all our modern needs occurs by transforming various types of energy into electrical power. Often it is mechanical energy that is transformed. This is done through the use of turbines, which are rotary engines that extract energy from fluid flow. A variety of power sources (energy types) may be used to drive these turbines including:

Hot gases or steam: Heat produced by nuclear reactions or the combustion of fossil fuels can be used to boil water and create steam. Alternatively, solar energy can be captured to heat the water. The gases themselves that result from the combustion of fossil fuels may also be used.

Water: As is seen in hydroelectric dams, the natural flow of water can be used to drive a turbine.

Wind: Naturally occurring wind may be collected using windmills, which directly link to the turbine.

Other sources: Though too small to be useful for large-scale production of electricity, other types of mechanical energy, such as an internal combustion engine or even a person turning a hand crank or riding a bicycle can be used to power a turbine.

The turbines, in turn, are attached to electrical generators which actually convert the mechanical energy in the turbine into electricity. Most generators rely on electromagnetic induction; in a strong magnetic field, the winding motion generates an electrical current. Thus, the generator moves, but does not actually create electrical current. This is often compared to a water pump, since a pump similarly moves, but does not create water.

Finally, the electricity must be distributed to industrial, commercial, and residential facilities. Power is transferred from the power generation plant as **alternating current (AC)**. It is carried through overhead power lines, or occasionally underground lines in highly populated areas, at a voltage of 110 kV or above. This high voltage current is delivered to substations, where the voltage is lowered to 33 kV using a transformer. The electricity at this lower voltage is then distributed to end-users. The system of power transmission is referred to as a grid, though in fact there are many redundant pathways. This allows power generated at one station to be redirected if another station fails to generate sufficient electricity.

IV.   CHARACTERISTICS OF SCIENCE

**0017   UNDERSTAND THE CHARACTERISTICS OF SCIENTIFIC
         KNOWLEDGE AND THE PROCESS OF SCIENTIFIC INQUIRY**

**Skill 17.1   Demonstrating knowledge of the nature, purpose, and
                 characteristics of science (e.g., reliance on verifiable evidence)
                 and the limitations of science in terms of the kinds of
                 questions that can be answered**

Science may be defined as a body of knowledge that is systematically derived
from study, observations, and experimentation. Its goal is to <u>identify and
establish principles and theories</u> that may be applied to explain the phenomena
of our universe and to solve problems. Pseudoscience, on the other hand, is a
belief that is not warranted. There is no scientific methodology or application.
Some of the more classic examples of pseudoscience include witchcraft, alien
encounters or any topic that is explained by hearsay.

Only certain types of questions can truly be answered by science because the
scientific method relies on observable phenomena and the ability to repeat those
phenomena. That is, only hypotheses that can be *tested* are valid. Often this
means that we can <u>control the variables</u> in a system to an extent that allows us to
truly determine their effects. If we don't have full control over the variables, for
instance, in environmental biology, we can study several different naturally
occurring systems in which the desired variable is different.

Scientific research serves two purposes –
1. To investigate and acquire knowledge which is theoretical and
2. To do research which is of practical value.

Science is in a unique position to be able to serve humanity. Scientific research
comes from inquiry. An inquiring mind is thirsty, trying to find answers. The two
most important questions – why and how, are the starting points. A person who
is inquisitive asks questions and wants to <u>find out answers</u>.

Scientific research uses scientific method to methodically answer the questions.
Those who research follow the scientific method, which consists of a series of
logical steps designed to solve a problem or find answer to their problem.

The aim of the scientific method is to eliminate bias/prejudice from the
scientist/researcher. As human beings, we are influenced by our bias/prejudice
and this method helps to eliminate that. If all the steps of the scientific method
are followed as outlined, there is the maximum elimination of bias.

Scientific research is clearly different from the learning of other areas. Science demands evidence. Science requires experimenting to prove or disprove one's ideas or propositions. Science does not answer all our questions. Science doesn't say any thing about the cultural, moral and religious beliefs of the individuals. Nor is science an appropriate way of investigating religious truths; on the other hand, religion or philosophy are not appropriate methods of investigating scientific phenomena. It is up to us to use the information science provides and make our own decisions according to our beliefs and norms.

Science can give qualitative answers. The objective of qualitative data analysis is a complete and detailed description from which patterns or other useful information may be gained. For example, an ornithologist may observe individuals of a bird species over a period of time to identify the resources they need for food and shelter or to document mating behavior.

Science can also give quantitative answers. Analysis of quantitative data involves making detailed measurements, classifying features, county them and constructing statistical models in an attempt to explain observations. These can then be generalized to a larger population or the ideal case, and direct comparisons can be made between two data sets as long as valid dampling and significance techniques have been used.

Basically, quantitative research is objective; qualitative is subjective. Quantitative research seeks explanatory laws; qualitative research aims at in-depth description. Quantitative research measures what it assumes to be a static reality in hopes of developing universal laws and is well suited to establishing cause-and-effect relationships. Qualitative research is an exploration of what is assumed to be a dynamic reality. It does not claim that what is discovered in the process is universal, and thus necessarily replicable.

Skill 17.2 Recognizing the dynamic nature of scientific knowledge through the continual testing, revision, and occasional rejection of existing theories

Whether to choose a fundamentally quantitative or a qualitative design depends on the nature of the project, the type of information needed, the context of the study, and the availability of resources (time, money, and human). It is important to keep in mind that these are two different approaches, not necessarily polar opposites. In fact, elements of both designs can and should be used together in mixed-methods studies. In scientific disciplines such as chemistry, it is generally the norm to record at least some observations of both types.

Advantages of combining both types of data include:

1. Research development (one approach is used to inform the other, such as using qualitative research to develop an instrument to be used in quantitative research)

2. Increased validity (confirmation of results by means of different data types)

3. Complementarities (adding information, e.g., descriptions alongside measurements)

4. Providing additional resources for explaining unanticipated results or failed experiments.

The first step in scientific inquiry is posing a question to be answered. Next, a hypothesis is formed to provide a plausible explanation. An experiment is then proposed and performed to test this hypothesis. A comparison between the predicted and observed results is the next step. Conclusions are then formed and it is determined whether the hypothesis is correct or incorrect. If incorrect, the next step is to form a new hypothesis and the process is repeated.

Science is limited by the available technology. An example of this would be the relationship of the discovery of the cell and the invention of the microscope. As our technology improves, more hypotheses will become theories and possibly laws. Science is also limited by the data that is able to be collected. Data may be interpreted differently on different occasions. Science limitations cause explanations to be changeable as new technologies emerge.

Theories provide a framework to explain the known information of the time, but are subject to constant evaluation and updating. There is always the possibility that new evidence will conflict with a current theory. The development of the atomic theory is a good example of a theory that has been proposed and substantiated, then revised and refined as new techniques became available. See **Skill 11.1** for a discussion of the development of the current atomic theory. Some theories like the Theory of Spontaneous Generation and the Theory of Acquired Characteristics have been rejected as the concepts became better explained by newer knowledge. Other theories like the sun-centered solar system and the germ theory of disease that were initially rejected have become well-accepted over time due to advancements in science.

Skill 17.3   Determining an appropriate scientific hypothesis or investigative design for addressing a given problem

The scientific method is the basic process behind science. It involves several steps beginning with hypothesis formulation and working through to the conclusion.

Posing a question
Although many discoveries happen by chance, the standard thought process of a scientist begins with forming a question to research. The more limited the question, the easier it is to set up an experiment to answer it. Scientific questions frequently result from observation of events in nature or in the laboratory. An observation is not just a look at what happens. It also includes measurements and careful records of the event. Records could include photos, drawings, or written descriptions. The observations and data collection may provide answers, or they may lead to one or more questions.

Having arrived at a question, a scientist usually researches the scientific literature to see what is known about the question. Perhaps the question has already been answered, or another experimenter has found part of the solution. The scientist may want to test or reproduce the answer found in the literature. Or, the research might lead to a new question.

Developing a hypothesis
If the question has not yet been answered, the scientist may prepare for an experiment by making a hypothesis. A hypothesis is a statement of a possible answer to the question. It is a tentative explanation for a set of facts and can be tested by experiments. Although hypotheses are usually based on observations, they may also be based on a sudden idea or intuition or a mathematical theory. Once the question is formulated, the scientist will take an educated guess about the answer to the problem or question. This 'best guess' is the hypothesis.

Skill 17.4   Demonstrating knowledge of the principles and procedures for designing and carrying out scientific investigations (e.g., changing one variable at a time)

Conducting the test
To make a test fair, data from an experiment must have a **variable** or any condition that can be changed such as temperature or mass. A good test will try to manipulate as few variables as possible so as to see which variable is responsible for the result. This requires a second example of a **control**. A control is an extra setup in which all the conditions are the same except for the variable being tested.

Some experiments may test the effect of one thing on another under controlled conditions. Such experiments have two variables. The experimenter controls one variable, called the independent variable. The other variable, the dependent variable, shows the result of changing the independent variable.

For example, suppose a researcher wanted to test the effect of Vitamin A on the ability of rats to see in dim light. The independent variable would be the dose of Vitamin A added to the rats' diet. The dependent variable would be the intensity of light to which the rats respond. All other factors, such as time, temperature, age, water and other nutrients given to the rats, are held constant.

Chemists sometimes do short experiments "just to see what happens" or to see what a certain reaction produces. Often, these are not formal experiments. Rather they are ways of making additional observations about the behavior of matter.

Observe and record the data

In most experiments scientists collect quantitative data, which are data that can be measured with instruments. They also collect qualitative data, descriptive information from observations other than measurements. Interpreting data and analyzing observations are an important part of the scientific method. If data are not organized in a logical manner, incorrect conclusions can be drawn. Also, other scientists may not be able to follow or reproduce the results.

Reporting of the data should state specifics of how the measurements were calculated. A graduated cylinder needs to be read with proper procedures. As beginning students, technique must be part of the instructional process so as to give validity to the data.

Graphing data

Graphing utilizes numbers to demonstrate patterns. The patterns offer a visual representation, making it easier to draw conclusions.

Drawing a conclusion

After recording data, you compare your data with that of other groups. A conclusion is the judgment derived from the data results. A conclusion must address the hypothesis on which the experiment was based. The conclusions state whether or not the data support the hypothesis. If not, the conclusion should state what the experiment did show. If the hypothesis is not supported, the scientist uses the observations from the experiment to make a new or revised hypothesis and plan new experiments.

Normally, knowledge is integrated in the form of a **lab report**. A report has many sections. It should include a specific **title** and tell exactly what is being studied. The **abstract** is a summary of the report written at the beginning of the paper. The **purpose** should always be defined and will state the problem. The purpose should include the **hypothesis** (educated guess) of what is expected from the outcome of the experiment. The entire experiment should relate to this problem. The procedure used to obtain data is important to the outcome. Experiments consist of **controls** and **variables**. A control is the experiment run under normal conditions. The variable includes a factor that is changed. In biology, the variable may be light, temperature, pH, time, etc. The differences in tested variables may be used to make a prediction or form a hypothesis. Only one variable should be tested at a time. One would not alter both the temperature and pH of the experimental subject.

An **independent variable** is one that is changed or manipulated by the researcher. This could be the amount of light given to a plant or the temperature at which bacteria is grown. The **dependent variable** is that which is influenced by the independent variable.

Observations and **results** of the experiment should be recorded including all results from data. Drawings, graphs and illustrations should be included to support information. Observations are objective, whereas analysis and interpretation is subjective. A **conclusion** should explain why the results of the experiment either proved or disproved the hypothesis.

Skill 17.5   Recognizing the importance of and strategies for avoiding bias in scientific investigations

Scientific research can be biased in the choice of what data to consider, in the reporting or recording of the data, and/or in how the data are interpreted. The scientist's emphasis may be influenced by his/her nationality, sex, ethnic origin, age, or political convictions. For example, when studying a group of animals, male scientists may focus on the social behavior of the males and on typically male characteristics.

Although bias related to the investigator, the sample, the method, or the instrument may not be completely avoidable in every case, it is important to know the possible sources of bias and how bias could affect the evidence. Moreover, scientists need to be attentive to possible bias in their own work as well as that of other scientists.

Objectivity may not always be attained. However, one precaution that may be taken to guard against undetected bias is to have many different investigators or groups of investigators working on a project. By different, it is meant that the groups are made up of various nationalities, ethnic origins, ages, and political convictions and composed of both males and females. It is also important to note one's aspirations, and to make sure to be truthful to the data, even when livelihoods, grants, promotions, and notoriety are at risk.

Another area of bias comes from poor instrument calibration, interference from other compounds, or loss of substance during sample pre-processing.

In order to avoid bias, it is imperative to conduct each experiment under exactly the same conditions, including a *control* experiment with a known negative outcome. Additionally, in order to avoid experimental bias, a researcher must not "read" particular results into data.

An example of experimental bias is as follows. A researcher is timing mice as they move through a maze toward a piece of cheese. The experiment relies on the mouse's ability to smell the cheese as it approaches. If one mouse chases a piece of Cheddar cheese, while another chases Limburger (a cheese with a very strong odor), the Limburger mouse may have a large advantage over the Cheddar mouse because it can smell the cheese more easily. To remove the experimental bias from this experiment, the same cheese should be used for both mice.

Scientists are expected to truthfully report the results of their experiments without fabricating data. Falsification of certain kinds of results has become easier with the advent of computer software that allows images to be manipulated. An error in one person's research may lead to incorrect conclusions that affect other scientists' research, and each scientist must accept a share of the responsibility for ensuring that the collective body of knowledge is accurate. For example, it was recently discovered that a computer model being used by one laboratory had transposed the arrangement of atoms in molecular structures to their mirror images. The published structures had already been relied upon by other researchers in the field for three years, and now many papers are having to be retracted and years of work redone. Had the original laboratory checked their computer model more carefully, this situation could have been avoided.

0018 UNDERSTAND SCIENTIFIC TOOLS, INSTRUMENTS, MATERIALS, AND SAFETY PRACTICES

Skill 18.1 Recognizing procedures for the safe and proper use of scientific tools, instruments, chemicals, and other materials in investigations

Bunsen burners - Hot plates should be used whenever possible to avoid the risk of burns or fire. If Bunsen burners are used, the following precautions should be followed:

1. Know the location of fire extinguishers and safety blankets and train students in their use. Long hair and long sleeves should be secured and out of the way.

2. Turn the gas all the way on and make a spark with the striker. The preferred method to light burners is to use strikers rather than matches.

3. Adjust the air valve at the bottom of the Bunsen burner until the flame shows an inner cone.

4. Adjust the flow of gas to the desired flame height by using the adjustment valve.

5. Do not touch the barrel of the burner (it is hot).

Light microscopes are commonly used in laboratory experiments. Several procedures should be followed to properly care for this equipment:

- Clean all lenses with lens paper only.
- Carry microscopes with two hands; one on the arm and one on the base.
- Always begin focusing on low power, then switch to high power.
- Store microscopes with the low power objective down.
- Always use a coverslip when viewing wet mount slides.
- Bring the objective down to its lowest position then focus by moving up to avoid breaking the slide or scratching the lens.

Wet mount slides should be made by placing a drop of water on the specimen and then putting a glass coverslip on top of the drop of water. Dropping the coverslip at a forty-five degree angle will help in avoiding air bubbles. Total magnification is determined by multiplying the ocular (usually 10X) and the objective (usually 10X on low, 40X on high).

Chemical purchase, use, and disposal

- Inventory all chemicals on hand at least annually. Keep the list up-to-date as chemicals are consumed and replacement chemicals are received.
- If possible, limit the purchase of chemicals to quantities that will be consumed within one year and that are packaged in small containers suitable for direct use in the lab without transfer to other containers.
- Label all chemicals to be stored with date of receipt or preparation and have labels initialed by the person responsible.
- Generally, bottles of chemicals should not remain:
  - Unused on shelves in the lab for more than one week. Move these chemicals to the storeroom or main stockroom.
  - Unused in the storeroom near the lab for more than one month. Move these chemicals to the main stockroom.
- Check shelf life of chemicals. Properly dispose of any out-dated chemicals.
- Ensure that the disposal procedures for waste chemicals conform to environmental protection requirements.
- Do not purchase or store large quantities of flammable liquids. Fire department officials can recommend the maximum quantities that may be kept on hand.
- Never open a chemical container until you understand the label and the relevant portions of the MSDS.

Chemical storage plan for laboratories

- Chemicals should be stored according to hazard class (ex. flammables, oxidizers, health hazards/toxins, corrosives, etc.).

- Store chemicals away from direct sunlight or localized heat.

- All chemical containers should be properly labeled, dated upon receipt, and dated upon opening.

- Store hazardous chemicals below shoulder height of the shortest person working in the lab.

- Shelves should be painted or covered with chemical-resistant paint or chemical-resistant coating.

- Shelves should be secure and strong enough to hold chemicals being stored on them. Do not overload shelves.

- Personnel should be aware of the hazards associated with all hazardous materials.

- Separate solids from liquids.

- Store each of the following in their own areas: acids, bases, flammables, peroxide-forming chemicals, water-reactive chemicals, oxidizers, and toxins

- A "Poison Control Network" telephone number should be posted in the laboratory where toxins are stored. Color-coded labeling systems that may be found in your lab are shown below:

Hazard	Color Code
Flammables	Red
Health Hazards/Toxins	Blue
Reactives/Oxidizers	Yellow
Contact Hazards	White
General Storage	Gray, Green, Orange

Please Note: Chemicals with labels that are colored and striped may react with other chemicals in the same hazard class. See MSDS for more information. Chemical containers which are not color-coded should have hazard information on the label. Read the label carefully and store accordingly.

Disposal of chemical waste

Schools are regulated by the Environmental Protection Agency, as well as state and local agencies, when it comes to disposing of chemical waste. Check with your state science supervisor, local college or university environmental health and safety specialists, and the Laboratory Safety Workshop as well as your school system for advice on the disposal of chemical waste. The American Chemical Society publishes an excellent guidebook, *Laboratory Waste Management, A Guidebook* (1994).

Chemical hazard pictorial

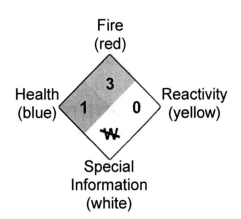

Fire
(red)

Health
(blue)

Reactivity
(yellow)

Special
Information
(white)

Several different pictorials are used on labels to indicate the level of a chemical hazard. The most common is the **"fire diamond" NFPA (National Fire Prevention Association) pictorial** shown below. A zero indicates a minimal hazard and a four indicates a severe risk. Special information includes if the chemical reacts with water, **OX** for an oxidizer, **COR ACID** for a corrosive acid, and **COR ALK** for a corrosive base. The "Health" hazard level is for **acute toxicity only**.

Pictorials are designed for **quick reference in emergency situations**, but they are also useful as minimal summaries of safety information for a chemical. They are not required on chemicals you purchase, so it's a good idea to add a label pictorial to every chemical you receive if one is not already present. **The entrance to areas where chemicals are stored should carry a fire diamond label** to represent the materials present.

Procedures for flammable materials: minimize fire risk

The vapors of a flammable liquid **or solid** may travel across the room to an ignition source and cause a fire or explosion.

- Store flammables in an approved safety cabinet. Store in safety cans if possible.
- Minimize volumes and concentrations used in an experiment with flammables.
- Minimize the time containers are open.
- Minimize ignition sources in the laboratory.
- Ensure that there is good air movement in the laboratory before the experiment.
- Check the fire extinguishers and be certain that you know how to use them.
- Tell the students that the "Stop, Drop, and Roll" technique is best for a clothing fire outside the lab, but in the lab they should walk calmly to the safety shower and use it. Practice this procedure with students in drills.
- A fire blanket should not be used for clothing fires because clothes often contain polymers that melt onto the skin. Pressing these fabrics into the skin with a blanket increases burn damage.
- If a demonstration of an exploding gas or vapor is performed, it should be done behind a safety shield using glass vessels taped with fabric tape.

Fitting and cleaning glassware

If a thermometer, glass tube, or funnel must be threaded through a stopper or a piece of tubing and it won't fit, either make the hole larger or use a smaller piece of glass. Use soapy water or glycerol to lubricate the glass before inserting it. Hold the glass piece as close as possible to the stopper during insertion. It's also good practice to wrap a towel around the glass and the stopper during this procedure. **Never apply undue pressure**.

Glassware sometimes contains **tapered ground-glass joints** to allow direct glass-to-glass connections. A thin layer of joint **grease** must be applied when assembling an apparatus with ground-glass joints. Too much grease will contaminate the experiment, and too little will cause the components to become tightly locked together. Disassemble the glassware with a **twisting** motion immediately after the experiment is over.

Cleaning glassware becomes more difficult with time, so it should be cleaned soon after the experiment is completed. Wipe off any lubricant with paper towel moistened in a solvent like hexane before washing the glassware. Use a brush with lab soap and water. Acetone may be used to dissolve most organic residues. Spent solvents should be transferred to a waste container for proper disposal.

Hot Plate

A **hot plate** (shown below) is used to heat Erlenmeyer flasks, beakers and other 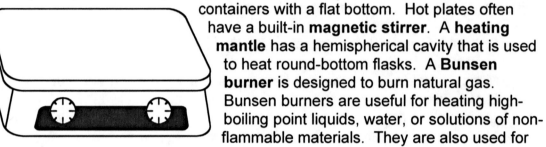 containers with a flat bottom. Hot plates often have a built-in **magnetic stirrer**. A **heating mantle** has a hemispherical cavity that is used to heat round-bottom flasks. A **Bunsen burner** is designed to burn natural gas. Bunsen burners are useful for heating high-boiling point liquids, water, or solutions of non-flammable materials. They are also used for bending glass tubing. Smooth boiling is achieved by adding **boiling stones** to a liquid.

Centrifuge

A **centrifuge** separates two immiscible phases by spinning the mixture (placed in a **centrifuge tube**) at high speeds. A **microfuge** or microcentrifuge is a small centrifuge. The weight of material placed in a centrifuge must be balanced, so if one sample is placed in a centrifuge, a tube with roughly an equal mass of water should be placed opposite the sample.

Skill 18.2   Identifying appropriate tools and units for measuring objects or substances

Science uses the metric system as it is accepted worldwide and allows easier comparison among experiments done by scientists around the world. Learn the following basic units and prefixes:

meter - measure of length
liter - measure of volume
gram - measure of mass

deca-(meter, liter, gram)= 10X the base unit   **deci** = 1/10 the base unit
hecto-(meter, liter, gram)= 100X the base unit   **centi** = 1/100 the base unit
kilo-(meter, liter, gram) = 1000X the base unit   **milli** = 1/1000 the base unit

Graduated Cylinder - These are used for precise measurements of **liquids**. They should always be placed on a flat surface. The surface of the liquid will form a meniscus (lens-shaped curve). The measurement is read at the <u>bottom</u> of this curve.

Balance - Electronic balances are easier to use, but more expensive. An electronic balance should always be tarred (returned to zero) before measuring and used on a flat surface. Substances should always be placed on a piece of paper to avoid spills and/or damage to the instrument. Triple beam balances must be used on a level surface. There are screws located at the bottom of the balance to make any adjustments. Start with the largest counterweight first and proceed toward the last notch that does not tip the balance. Do the same with the next largest, etc until the pointer remains at zero. The total **mass** is the total of all the readings on the beams. Again, use paper under the substance to protect the equipment.

Buret – A buret is used to dispense precisely measured volumes of liquid. A stopcock is used to control the volume of liquid being dispensed at a time.

Thermometers – The Celsius scale is designed to have 100 degrees between the boiling point and freezing point of water at standard atmospheric pressure. The Kelvin scale is designed so that no negative numbers are used since it starts at absolute zero. At sea level, water freezes at around 273 K and boils at 373 K. Notice that there are still 100 degrees between the freezing and boiling point temperatures on the Kelvin scale. Temperatures on this scale are called **kelvins**, *not* degrees kelvin, kelvin is *not* capitalized, and the symbol (capital K) stands alone with no degree symbol.

Skill 18.3 Identifying potential safety hazards associated with scientific equipment, materials, procedures, and settings

Safety in the science classroom and laboratory is of paramount importance to the science educator. The following is a general summary of the types of safety equipment that should be made available within a given school system as well as general locations where the protective equipment or devices should be maintained and used. Please note that this is only a partial list and that your school system should be reviewed for unique hazards and site-specific hazards at each facility.

The key to maintaining a safe learning environment is through proactive training and regular in-service updates for all staff and students who utilize the science laboratory. Proactive training should include how to **identify potential hazards**, **evaluate potential hazards**, and **how to prevent or respond to hazards**.

The following types of training should be considered:

a) Right to Know (OSHA training on the importance and benefits of properly recognizing and safely working with hazardous materials) along with some basic chemical hygiene as well as how to read and understand a material safety data sheet,
b) instruction in how to use a fire extinguisher,
c) instruction in how to use a chemical fume hood,
d) general guidance in when and how to use personal protective equipment (e.g. safety glasses or gloves), and
e) instruction in how to monitor activities for potential impacts on indoor air quality.

It is also important for the instructor to utilize **Material Data Safety Sheets**. Maintain a copy of the material safety data sheet for every item in your chemical inventory. This information will assist you in determining how to store and handle your materials by outlining the health and safety hazards posed by the substance. In most cases the manufacturer will provide recommendations with regard to protective equipment, ventilation and storage practices. This information should be your first guide when considering the use of a new material.

Frequent monitoring and in-service training on all equipment, materials, and procedures will help to ensure a safe and orderly laboratory environment. It will also provide everyone who uses the laboratory the safety fundamentals necessary to discern a safety hazard and to respond appropriately.

Work habits
- Never work alone in a laboratory or storage area.
- Never eat, drink, smoke, apply cosmetics, chew gum or tobacco, or store food or beverages in a laboratory environment or storage area.
- Keep containers closed when they are not in use.
- Never pipet by mouth.
- Restrain loose clothing and long hair and remove dangling jewelry.
- Tape all Dewar flasks with fabric-based tape.
- Check all glassware before use. Discard it if chips or star cracks are present.
- Never leave heat sources unattended.
- Do not store chemicals and/or apparatus on the lab bench or on the floor or aisles of the lab or storage room.
- Keep lab shelves organized.
- Never place a chemical, not even water, near the edges of a lab bench.
- Use a fume hood that is known to be in operating condition when working with toxic, flammable, and/or volatile substances.
- Never put your head inside a fume hood.
- Never store anything in a fume hood.

- Obtain, read, and be sure you understand the MSDS (see below) for each chemical that is to be used before allowing students to begin an experiment.
- Analyze new lab procedures and student-designed lab procedures in advance to identify any hazardous aspects. Minimize and/or eliminate these components before proceeding. Ask yourself these questions:
  - What are the hazards?
  - What are the worst possible things that could go wrong?
  - How will I deal with them?
  - What are the prudent practices, protective facilities and equipment necessary to minimize the risk of exposure to the hazards?
- Analyze close calls and accidents to eliminate their causes and prevent them from occurring again.
- Identify which chemicals may be disposed of in the drain by consulting the MSDS or the supplier. Clear one chemical down the drain by flushing with water before introducing the next chemical.
- Preplan for emergencies:
  - Keep the fire department informed of your chemical inventory and its location.
  - Consult with a local physician about toxins used in the lab and ensure that your area is prepared in advance to treat victims of toxic exposure.
  - Identify devices that should be shut off if possible in an emergency.
  - Inform your students of the designated escape route and alternate route.

Skill 18.4 Demonstrating knowledge of procedures for the ethical use and care of living organisms in scientific research

Dissections - Animals that are not obtained from recognized sources should not be used. Decaying animals or those of unknown origin may harbor pathogens and/or parasites. Specimens should be rinsed before handling. Latex gloves are desirable. If gloves are not available, students with sores or scratches should be excused from the activity. Formaldehyde is a carcinogen and should be avoided or disposed of according to district regulations. Students objecting to dissections for moral reasons should be given an alternative assignment.

Live specimens - No dissections may be performed on living mammalian vertebrates or birds. Lower order life and invertebrates may be used. Biological experiments may be done with all animals except mammalian vertebrates or birds. No physiological harm may result to the animal. All animals housed and cared for in the school must be handled in a safe and humane manner. Animals are not to remain on school premises during extended vacations unless adequate care is provided. Many state laws stipulate that any instructor who intentionally refuses to comply with the laws may be suspended or dismissed.

Microbiology - Pathogenic organisms must never be used for experimentation. Students should adhere to the following rules at all times when working with microorganisms to avoid accidental contamination:

1. Treat all microorganisms as if they were pathogenic.
2. Maintain sterile conditions at all times

If you are taking a national level exam you should check the Department of Education for your state for safety procedures. You will want to know what your state expects of you not only for the test but also for performance in the classroom and for the welfare of your students.

Skill 18.5 Recognizing appropriate protocols for maintaining safety and for responding to emergencies during classroom laboratory activities

All science labs should contain the following items of safety equipment. The following are requirements by law.

- Fire blanket which is visible and accessible
- Ground Fault Circuit Interrupters (GFCI) within two feet of water supplies
- Emergency shower capable of providing a continuous flow of water
- Signs designating room exits
- Emergency eye wash station which can be activated by the foot or forearm
- Eye protection for every student and a means of sanitizing equipment
- Emergency exhaust fans providing ventilation to the outside of the building
- Master cut-off switches for gas, electric, and compressed air. Switches must have permanently attached handles. Cut-off switches must be clearly labeled.
- An ABC fire extinguisher
- Storage cabinets for flammable materials

Also recommended, but not required by law:

- Chemical spill control kit
- Fume hood with a motor which is spark proof
- Protective laboratory aprons made of flame retardant material
- Signs which will alert people to potential hazardous conditions
- Containers for broken glassware, flammables, corrosives, and waste.
- Containers should be labeled.

It is the responsibility of teachers to provide a safe environment for their students. Proper supervision greatly reduces the risk of injury and a teacher should never leave a class for any reason without providing alternate supervision. After an accident, two factors are considered; foreseeability and negligence. **Foreseeability** is the anticipation that an event may occur under certain circumstances. **Negligence** is the failure to exercise ordinary or reasonable care. Safety procedures should be a part of the science curriculum and a well-managed classroom is important to avoid potential lawsuits

The **"Right to Know Law" statutes** cover science teachers who work with potentially hazardous chemicals. Briefly, the law states that employees must be informed of potentially toxic chemicals. An inventory must be made available if requested. The inventory must contain information about the hazards and properties of the chemicals. Training must be provided in the safe handling and interpretation of the Material Safety Data Sheet.

The following chemicals are potential carcinogens and are not allowed in school facilities:

> Acrylonitriel, Arsenic compounds, Asbestos, Bensidine, Benzene, Cadmium compounds, Chloroform, Chromium compounds, Ethylene oxide, Ortho-toluidine, Nickel powder, Mercury.

All laboratory solutions should be prepared as directed in the lab manual. Care should be taken to avoid contamination. All glassware should be rinsed thoroughly with distilled water before using, and cleaned well after use. Safety goggles should be worn while working with glassware in case of an accident. All solutions should be made with distilled water as tap water contains dissolved particles, which may affect the results of an experiment. Chemical storage should be located in a secured, dry area. Chemicals should be stored in accordance with reactivity. Acids are to be locked in a separate area. Used solutions should be disposed of according to local disposal procedures. Any questions regarding safe disposal or chemical safety may be directed to the local fire department.

0019 UNDERSTAND SCIENTIFIC COMMUNICATION AND THE SKILLS AND PROCEDURES FOR ANALYZING DATA

Skill 19.1 Recognizing the concepts of precision, accuracy, and error and identifying potential sources of error in gathering and recording data

Accuracy and precision

Accuracy is the degree of conformity of a measured quantity to its actual (true) value. A measurement is **accurate** when it agrees with the true value of the quantity being measured. An accurate measurement is **valid**. We get the right answer.

A measurement is **precise** when individual measurements of the same quantity agree with one another. Precision is also called reproducibility or repeatability and is the degree to which other measurements will show the same or similar results. A precise measurement is **reproducible**. We get a similar answer each time.

Accuracy is the degree of veracity while precision is the degree of reproducibility. These terms are related to **sources of error** in a measurement.
The best analogy to explain accuracy and precision is the target comparison:

- Repeated measurements are compared to arrows that are fired at a target.
- Accuracy describes the closeness of arrows to the bull's eye at the target center.
- Arrows that strike closer to the bull's eye are considered more accurate.
- Arrows that strike closer to each other are considered more precise.
- It is possible to be accurate without being precise; it is also possible to be precise without being accurate.

Systematic and random error

All experimental uncertainty is due to either random errors or systematic errors.

Random error results from **limitations in equipment or techniques**. **Random error decreases precision**. Remember that all measurements reported to a proper number of significant digits contain an imprecise final digit to reflect random error.

Systematic error results from **imperfect equipment or technique**. **Systematic error decreases accuracy**. Instead of a random error with random fluctuations, there is a biased result that on average is too large or small.

Systematic errors are reproducible inaccuracies that are consistently in the same direction. Systematic errors are often due to a problem, which persists throughout the entire experiment.

Systematic and random errors refer to problems associated with making measurements. Mistakes made in the calculations or in reading the instrument are not considered in error analysis.

Skill 19.2   Applying appropriate mathematical concepts and computational skills to analyze data (e.g., using ratios; determining mean, median, and mode)

Modern science utilizes a number of other disciplines. Statistics is one of those subjects, and is essential for understanding science. Statistical measures are employed to characterize trends in data.

Mean: Mean is the mathematical average of all the items. To calculate the mean, all the items must be added up and divided by the number of items. This is also called the **arithmetic mean** or more commonly as the "average". The arithmetic mean is a good measure of the central tendency of roughly normal distributions, but may be misleading in skewed distributions. In cases of skewed distributions, other statistics such as the median or geometric mean may be more informative. The **geometric mean** is found by multiplying all the values together, then taking the nth root of the result, where n is the number of data points.

Median: The median is the middle of a distribution: half the scores are above the median and half are below it. Unlike the mean, the median is not highly sensitive to extreme data points. This makes the median a better measure than the mean for finding the central tendency of highly skewed distributions. The median is determined by organizing the data points from lowest to highest. When there is an odd number of numbers, the median is simply the middle number. For example, the median of 2, 4, and 7 is 4. When there is an even number of numbers, the median is the mean of the two middle numbers. Thus, the median of the numbers 2, 4, 7, and 12 is (4+7)/2 = 5.5.

Mode: Mode is the value of the item, or data point, that occurs the most often in the distribution and is used as a measure of central tendency. Bimodal is a situation where there are two items with equal frequency. In the case of a perfectly normal distribution, the mean, median, and mode are identical.

Range: Range is the difference between the maximum and minimum values. The range is the difference between two extreme points on the distribution curve.

Skill 19.3   Identifying methods (e.g., tables, graphs) and criteria for organizing data to aid in the analysis of data (e.g., detecting patterns)

Classifying is grouping items according to their similarities. It is important for students to realize relationships and similarity as well as differences to reach a reasonable conclusion in a lab experience.

Graphing is an important skill to visually display collected data for analysis. The type of graphic representation used to display observations depends on the data that is collected. There are seven basic types of graphs and charts: column graphs, line graphs, bar graphs, scatter plots, pie charts, and area charts. The interpretation of data and the construction and interpretation of graphs are central practices in science. Graphs are effective visual tools which relay information quickly and reveal trends easily. While there are several different types of graphical displays, extracting information from them can be described in three basic steps.

1. Describe the graph: What does the title say? What is displayed on the x- and y-axes, including the units? Notice any symbols used and check for a legend or explanation.

2. Describe the data: Identify the range of data. Are patterns reflected in the data?

3. Interpret the data: How do patterns seen in the graph relate to other variables? What conclusions can be drawn from the patterns?

Line graphs are used to compare different sets of related data or to predict data that has not yet be measured. An example of a line graph would be comparing the rate of activity of different enzymes at varying temperatures.

Line graphs are set up to show two variables represented by one point on the graph. The X axis is the horizontal axis and represents the dependent variable. Dependent variables are those that would be present independently of the experiment. A common example of a dependent variable is time. Time proceeds regardless of anything else that occurs. The Y axis is the vertical axis and represents the independent variable. Independent variables are manipulated by the experiment, such as the amount of

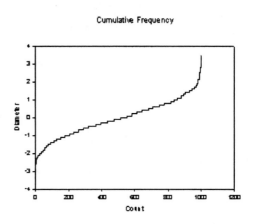

light, or the height of a plant. Graphs should be calibrated at equal intervals. If one space represents one day, the next space may not represent ten days. A "best fit" line is drawn to join the points and may not include all the points in the data. Axes must always be labeled for the graph to be meaningful. A good title will describe both the dependent and the independent variable.

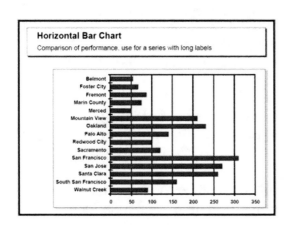

A **bar graph** or **histogram** is used to compare different items and make comparisons based on this data. An example of a bar graph would be comparing the ages of children in a classroom. Bar graphs are set up similarly in regards to axes, but points are not plotted. Instead, the dependent variable is set up as a bar where the X axis intersects with the Y axis. Each bar is a separate item of data and is not joined by a continuous line.

A **pie chart** is a circle with radii connecting the center to the edge. The area between two radii is called a slice. Data values are proportionate to the angle between the radii.

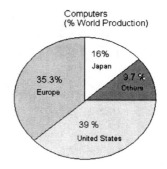

A pie chart is useful when organizing data as part of a whole. A good use for a pie chart would be displaying the percent of time students spend on various after school activities. Be careful not to include too many slices, as this may result in a cluttered graph. Six slices are typically as many as can be handled on one pie.

As noted before, the experimenter controls the independent variable. This variable is placed on the x-axis (horizontal axis). The dependent variable is influenced by the independent variable and is placed on the y-axis (vertical axis). It is important to choose the appropriate units for labeling the axes. It is best to take the largest value to be plotted and divide it by the number of block, and rounding to the nearest whole number.

Skill 19.4 Demonstrating knowledge of the use of data to support or challenge scientific arguments and claims

Because people often attempt to use scientific evidence in support of political or personal agendas, the ability to evaluate the credibility of scientific claims is a necessary skill in today's society. In evaluating scientific claims made in the media, public debates, and advertising, one should follow several guidelines.

First, scientific, peer-reviewed journals are the most accepted source for information on scientific experiments and studies. One should carefully scrutinize any claim that does not reference peer-reviewed literature.

Second, the media and those with an agenda to advance (advertisers, debaters, etc.) often overemphasize the certainty and importance of experimental results. One should question any scientific claim that sounds fantastical or overly certain.

Finally, knowledge of experimental design and the scientific method is important in evaluating the credibility of studies. For example, one should look for the inclusion of control groups and the presence of data to support the given conclusions.

Skill 19.5 Identifying appropriate methods for communicating the outcomes of scientific investigations (e.g., publication in peer-reviewed journals)

Conclusions must be communicated by clearly describing the information using accurate data, visual presentation and other appropriate media such as a slide (PowerPoint) presentation. Examples of visual presentations are graphs (bar/line/pie), tables/charts, diagrams, and artwork. Modern technology must be used whenever necessary. The method of communication must be suitable to the audience. Written communication is as important as oral communication. The scientist's strongest ally is a solid set of reproducible data.

Peer review is the process by which scientific results produced by one group are subjected to the analysis of other experts in the field. In practice it is most often used by scientific journals. Scientists author manuscripts detailing their experiments, results, and interpretations and these manuscripts are distributed by the journal editors to other researchers in the field for review prior to publication. The authors must address the comments and questions of the reviewers and make appropriate revisions for their work to be accepted for publication. Peer review is also the process by which applications for research funds are evaluated and awarded. Peer review may also be used informally by groups of researchers or graduate students wishing to get an evaluation of their research prior to writing it up for publication.

Reviewers of scientific work are typically experts in the field, but it is important that they be objective in their evaluations because it is possible that the results under review may contradict the ideas of the reviewers. Peer review is typically done **anonymously** so that the identities of the reviewers remain unknown by the scientists submitting work for review. However, less formal peer review may occur through lunch seminars, presentations at scientific conferences, and other venues where comments and responses may be provided in person.

The goal of peer review is to "weed out" science not performed to appropriate standards. This typically means the scientific method has been employed but also state of the art technical procedures have been followed and the conclusions that are drawn are fully supported by the results. Therefore, the reviewer will determine whether proper controls were in place, enough replicates were performed, and that the experiments clearly address the presented hypothesis. The reviewer will scrutinize the interpretations and how they fit into what is already known in the field. Often reviewers will suggest that additional experiments be done to further corroborate presented conclusions.

Occasionally, new scientific results may contradict long held ideas in a particular field. In these cases, in-depth and objective peer review is highly important. Scientists must work together to determine whether the new evidence is correct and how it might change current theories. Typically, there is resistance to the overthrow of scientific theories and many, many experiments must be arduously validated before new, contradictory hypotheses are accepted. Unfortunately, scientists are still people and so can be stubborn and slow to change their ideas. Therefore, **acceptance of new scientific results, even when experiments have been correctly performed, often takes some time.**

Skill 19.6 **Demonstrating familiarity with effective and valid resources, applying strategies for reading to gain information about science-related topics, and developing subject-area vocabulary**

<u>**Science Reference Resources**</u>:

Encyclopedia Britannica, Wikipedia, PubMed, Encarta, All Science Texts

List of Science Journals <u>http://www.loc.gov/rr/scitech/cjoulist.html</u>

<u>**Strategic Reading Resources**</u>:

Language Arts – Reading Strategies – What are methods for teaching reading strategies? <u>http://www.howard.k12.md.us/langarts/Curriculum/strategies.htm</u>

Reading Strategies – Scaffolding Students' Interaction with Text <u>http://www.greece.k12.ny.us/instruction/ela/6-12/Reading/Reading%20Strategies/reading%20strategies%20index.htm</u>

Mind Tools – Reading Strategies <u>http://www.mindtools.com/rdstratg.html</u>

<u>**Strategies**</u>:

To optimize the amount of information and vocabulary retained by the reading student, teachers must encourage strategic reading methods. Strategic reading is the process whereby students actively construct meaning and interact with text as they read by setting purposes for reading, establishing methods of accomplishing such purposes, monitoring comprehension as they read, and evaluating the completed task.

There are several steps a strategic reader must follow before, during and after reading. Before reading, the student must build up background knowledge on the reading and the topic, set a purpose for the reading, and determine methods to accomplish this purpose. While reading, students must devote their entire attention to the task, continually confirm their understanding, use semantic and syntactic cues to construct meanings of unfamiliar words, use outside sources to define unknown words, and ask questions. After reading, strategic readers must decide if their goals for reading have been reached, evaluate their understanding of reading material, summarize major ideas, construct lists of new vocabulary, seek any additional needed information from outside sources and paraphrase what they have learned.

Teachers of strategic reading should follow several clearly defined steps. First, the teacher must introduce the strategic reading method by discussing reasons why it is taught and explaining the steps involved. Then, the teacher must provide the opportunity for collaborative group work and individual work with the strategy. After this session, the teacher should discuss with the students what was done with the strategy and why. Finally, the steps of the strategy should be reviewed again, and the strategy should be referenced and revisited whenever possible.

Prior to reading, strategies should stress the recall of prior knowledge, prediction of what reading will teach and the selection of reading materials to suit their purpose. To teach before-reading strategies, teachers can employ "think-a-louds," a type of verbal brainstorming, and "previewing," where teachers express prior personal experiences relevant to the text and encourage students to do so while reading background information about the author.

During-reading activities should enable students to evaluate their own comprehension while reading and to improve comprehension if needed. One such method is self-monitoring, which is an active awareness that strategic readers have of their control over their own understanding. Self-questioning is another during-reading strategy in which students generate questions about important elements of the text and new vocabulary which are used to retain information and also for group discussion when reading has been completed.

After-reading strategies help to increase comprehension and retained information. These strategies include summarizing what has been read, interpreting and evaluating ideas, applying these ideas to situations outside of the reading, creating lists of new terms and ideas, and using study strategies for note taking and locating important content.

0020 UNDERSTAND THE UNIFYING CONCEPTS OF SCIENCE AND TECHNOLOGY

Skill 20.1   Demonstrating knowledge of the unifying concepts (e.g., system, model, change, scale) of science and technology

Math, science, and technology have common themes in how they are applied and understood. All three use models, diagrams, and graphs to simplify a concept for analysis and interpretation. Patterns observed in these systems lead to predictions based on these observations. The following are the concepts and processes generally recognized as common to all scientific disciplines:

- Systems, order, and organization

- Evidence, models, and explanation

- Constancy, change, and measurement

- Evolution and equilibrium

- Form and function

Systems, order, and organization

Because the natural world is so complex, the study of science involves the **organization** of items into smaller groups based on interaction or interdependence. These groups are called **systems**. Examples of organization are the periodic table of elements and the five-kingdom classification scheme for living organisms. Examples of systems are the solar system, cardiovascular system, Newton's laws of force and motion, and the laws of conservation.

Order refers to the behavior and measurability of organisms and events in nature. The arrangement of planets in the solar system and the life cycle of bacterial cells are examples of order.

Evidence, models, and explanations

Scientists use **evidence** and **models** to form **explanations** of natural events. Models are miniaturized representations of a larger event or system. Evidence is anything that furnishes proof.

Constancy, change, and measurement

Constancy and **change** describe the observable properties of natural organisms and events. Scientists use different systems of **measurement** to observe change and constancy. For example, the freezing and melting points of given substances and the speed of sound are constant under constant conditions. Growth, decay, and erosion are all examples of natural change.

Evolution and equilibrium

Evolution is the process of change over a long period of time. While biological evolution is the most common example, one can also classify technological advancement, changes in the universe, and changes in the environment as evolution.

Equilibrium is the state of balance between opposing forces of change. Homeostasis and ecological balance are examples of equilibrium.

Form and function

Form and **function** are properties of organisms and systems that are closely related. The function of an object usually dictates its form and the form of an object usually facilitates its function. For example, the form of the heart (e.g. muscle, valves) allows it to perform its function of circulating blood through the body.

Skill 20.2   Recognizing the characteristics of systems, how the components of a system interact (e.g., negative and positive feedback), and how different systems interact with one another

Because the natural world is so complex, the study of science involves the **organization** of items into smaller groups based on interaction or interdependence. These groups are called **systems**. Examples of organization are the periodic table of elements and the five-kingdom classification scheme for living organisms. Examples of systems are the solar system, cardiovascular system, Newton's laws of force and motion, and the laws of conservation.

In animal and human systems, feedback loops serve to regulate bodily functions in relation to environmental conditions. Positive feedback loops enhance the body's response to external stimuli and promote processes that involve rapid deviation from the initial state. For example, positive feedback loops function in stress response and the regulation of growth and development. Negative feedback loops help maintain stability in spite of environmental changes and function in homeostasis. For example, negative feedback loops function in the regulation of blood glucose levels and the maintenance of body temperature.

Feedback loops regulate the secretion of classical vertebrate hormones in humans. The pituitary gland and hypothalamus respond to varying levels of hormones by increasing or decreasing production and secretion. High levels of a hormone cause down-regulation of the production and secretion pathways, while low levels of a hormone cause up-regulation of the production and secretion pathways.

"Fight or flight" refers to the human body's response to stress or danger. Briefly, as a response to an environmental stressor, the hypothalamus releases a hormone that acts on the pituitary gland, triggering the release of another hormone, adrenocorticotropin (ACTH), into the bloodstream. ACTH then signals the adrenal glands to release the hormones cortisol, epinephrine, and norepinephrine. These three hormones act to ready the body to respond to a threat by increasing blood pressure and heart rate, speeding reaction time, diverting blood to the muscles, and releasing glucose for use by the muscles and brain. The stress-response hormones also down-regulate growth, development, and other non-essential functions. Finally, cortisol completes the "fight or flight" feedback loop by acting on the hypothalamus to stop hormonal production after the threat has passed.

Skill 20.3   Identifying types and characteristics of models used in science and technology and the advantages and limitations of models

The model is a basic adjunct to the scientific method. Many things in science are studied with models. A model is any simplification or substitute for what we are actually studying, understanding or predicting. A model is a substitute, but it is similar to what it represents. We encounter models at every step of our daily living. The Periodic Table of the elements is a model chemists use for predicting the properties of the elements. Physicists use Newton's laws to predict how objects will interact, such as planets and spaceships. In geology, the continental drift model predicts the past positions of continents. Sample, ideas, and methods are all examples of models. At every step of scientific study models are extensively use. The primary activity of the hundreds of thousands of US scientists is to produce new models, resulting in tens of thousands of scientific papers published per year.

Types of models:

- Scale models: some models are basically downsized or enlarged copies of their target systems like the models of protein, DNA etc.
- idealized models: An idealization is a deliberate simplification of something complicated with the objective of making it easier to understand. Some examples are frictionless planes, point masses, isolated systems etc.
- Analogical models: standard examples of analogical models are the billiard model of a gas, the computer model of the mind, or the liquid drop model of the nucleus.
- Phenomenological models: These are usually defined as models that are independent of theories.
- Data models: It s a corrected, rectified, regimented, and in many instances, idealized
- version of the data we gain from immediate observation (raw data).

- Theory models: Any structure is a model if it represents an idea (theory). An example of this is a flow chart, which summarizes a set of ideas.

Uses of models:

1. Models are crucial for understanding the structure and function of processes in science.
2. Models help us to visualize the organs/systems they represent just like putting a face to person.
3. Models are very useful to predict and foresee future events like hurricanes etc.

Limitations of models:
1. Though models are every useful to us, they can never replace the real thing.
2. Models are not exactly like the real item they represent.
3. Caution must be exercised before presenting the models to the class, as they may not be accurate.
4. It is the responsibility of the educator to analyze the model critically for the proportions, content value, and other important data.
5. One must be careful about the representation style. This style differs from person to person.

Sample Test

Directions: Read each item and select the correct response. The answer key follows.

1. What is the main difference between the 'condensation hypothesis' and the 'tidal hypothesis' for the origin of the solar system?
(Rigorous) (Skill 1.1)

A.   The tidal hypothesis can be tested, but the condensation hypothesis cannot.

B.   The tidal hypothesis proposes a near collision of two stars pulling on each other, but the condensation hypothesis proposes condensation of rotating clouds of dust and gas.

C.   The tidal hypothesis explains how tides began on planets such as Earth, but the condensation hypothesis explains how water vapor became liquid on Earth.

D.   The tidal hypothesis is based on Aristotelian physics, but the condensation hypothesis is based on Newtonian mechanics.

2. Which of the following is the best definition for 'meteorite'?
(Average Rigor) (Skill 1.2)

A.   A meteorite is a mineral composed of mica and feldspar.

B.   A meteorite is material from outer space, that has struck the earth's surface.

C.   A meteorite is an element that has properties of both metals and nonmetals.

D.   A meteorite is a very small unit of length measurement.

3. The force of gravity on earth causes all bodies in free fall to _____ .

(Average Rigor) (Skill 1.3)

A.   fall at the same speed.

B.   accelerate at the same rate.

C.   reach the same terminal velocity.

D.   move in the same direction.

4. The electromagnetic radiation with the longest wave length is/are _____.
(Rigorous) (Skill 1.6)

A. radio waves.
B. red light.
C. X-rays.
D. ultraviolet light.

5. When heat is added to most solids, they expand. Why is this the case?
(Average Rigor) (Skill 2.1)

A. The molecules get bigger.
B. The faster molecular motion leads to greater distance between the molecules.
C. The molecules develop greater repelling electric forces.
D. The molecules form a more rigid structure.

6. Which of the following is *not* true about phase change in matter?
(Rigorous) (Skill 2.1)

A. Solid water and liquid ice can coexist at water's freezing point.
B. At 7 degrees Celsius, water is always in liquid phase.
C. Matter changes phase when enough energy is gained or lost.
D. Different phases of matter are characterized by differences in molecular motion.

7. What is the most accurate description of the Water Cycle?
(Easy) (Skill 2.2)

A. Rain comes from clouds, filling the ocean. The water then evaporates and becomes clouds again.
B. Water circulates from rivers into groundwater and back, while water vapor circulates in the atmosphere.
C. Water is conserved except for chemical or nuclear reactions, and any drop of water could circulate through clouds, rain, ground-water, and surface-water.
D. Weather systems cause chemical reactions to break water into its atoms.

8. What is the source for most of the United States' drinking water?
(Easy) (Skill 2.2)

A. Desalinated ocean water.
B. Surface water (lakes, streams, mountain runoff).
C. Rainfall into municipal reservoirs.
D. Groundwater.

9. The salinity of ocean water is closest to _____.
(Rigorous) (Skill 2.5)

A. 0.035 %
B. 0.35 %
C. 3.5 %
D. 35 %

10. The transfer of heat by electromagnetic waves is called _____.
(Easy) (Skill 3.2)

A. conduction.
B. convection.
C. phase change.
D. radiation.

11. Which of the following instruments measures wind speed? *(Rigorous) (Skill 13.7)*

A. A barometer.
B. An anemometer.
C. A wind sock.
D. A weather vane.

12. A cup of hot liquid and a cup of cold liquid are both sitting in a room at comfortable room temperature and humidity. Both cups are thin plastic. Which of the following is a true statement?
(Rigorous) (Skill 3.7)

A. There will be fog on the outside of the hot liquid cup, and also fog on the outside of the cold liquid cup.
B. There will be fog on the outside of the hot liquid cup, but not on the cold liquid cup.
C. There will be fog on the outside of the cold liquid cup, but not on the hot liquid cup.
D. There will not be fog on the outside of either cup.

13. Which of the following types of rock is made from magma?
(Average Rigor) (Skill 4.1)

A. Fossils
B. Sedimentary
C. Metamorphic
D. Igneous

14. Igneous rocks can be classified according to which of the following?
(Rigorous) (Skill 4.1)

A. Texture.
B. Composition.
C. Formation process.
D. All of the above.

15. Lithification refers to the process by which unconsolidated sediments are transformed into

_____.

(Rigorous) (Skill 4.2)

A.　metamorphic rocks.
B.　sedimentary rocks.
C.　igneous rocks.
D.　lithium oxide.

16. Which of these is a true statement about loamy soil?
(Average Rigor) (Skill 4.3)

A.　Loamy soil is gritty and porous.
B.　Loamy soil is smooth and a good barrier to water.
C.　Loamy soil is hostile to microorganisms.
D.　Loamy soil is velvety and clumpy.

17. _____are cracks in the plates of the earth's crust, along which the plates move.
(Easy) (Skill 4.5)

A.　Faults.
B.　Ridges.
C.　Earthquakes.
D.　Volcanoes.

18. Which of the following is *not* a type of volcano?
(Rigorous) (Skill 4.5)

A.　Shield Volcanoes.
B.　Composite Volcanoes.
C.　Stratus Volcanoes.
D.　Cinder Cone Volcanoes.

19. Fossils are usually found in _____ rock.
(Average Rigor) (Skill 4.6)

A.　igneous.
B.　sedimentary.
C.　metamorphic.
D.　cumulus.

20. Which of the following is the most accurate definition of a non-renewable resource?
(Easy) (Skill 5.1)

A.　A nonrenewable resource is never replaced once used.
B.　A nonrenewable resource is replaced on a timescale that is very long relative to human life-spans.
C.　A nonrenewable resource is a resource that can only be manufactured by humans.
D.　A nonrenewable resource is a species that has already become extinct.

21. Which of the following is *not* a common type of acid in 'acid rain' or acidified surface water?
(Rigorous) (Skill 5.4)

A.　Nitric acid.
B.　Sulfuric acid.
C.　Carbonic acid.
D.　Hydrofluoric acid.

22. Laboratory researchers have classified fungi as distinct from plants because the cell walls of fungi _____ .
(Average Rigor) (Skill 6.1)

A. contain chitin.
B. contain yeast.
C. are more solid.
D. are less solid.

23. Which kingdom is comprised of organisms made of one cell with no nuclear membrane?
(Rigorous) (Skill 6.1)

A. Monera.
B. Protista.
C. Fungi.
D. Algae.

24. Animals with a notochord or backbone are in the phylum
(Rigorous) (Skill 6.1)

A. arthropoda.
B. chordata.
C. mollusca.
D. mammalia.

25. The scientific name *Canis familiaris* refers to the animal's
_____.
(Rigorous) (Skill 6.3)

A. kingdom and phylum.
B. genus and species.
C. class and species.
D. type and family.

26. Members of the same animal species _____ .
(Average Rigor) (Skill 6.3)

A. look identical.
B. never adapt differently.
C. are able to reproduce with one another.
D. are found in the same location.

27. Which of the following is a correct explanation for scientific 'evolution'?
(Average Rigor) (Skill 6.5)

A. Giraffes need to reach higher for leaves to eat, so their necks stretch. The giraffe babies are then born with longer necks. Eventually, there are more long-necked giraffes in the population.
B. Giraffes with longer necks are able to reach more leaves, so they eat more and have more babies than other giraffes. Eventually, there are more long-necked giraffes in the population.
C. Giraffes want to reach higher for leaves to eat, so they release enzymes into their bloodstream, which in turn causes fetal development of longer-necked giraffes. Eventually, there are more long-necked giraffes in the population.
D. Giraffes with long necks are more attractive to other giraffes, so they get the best mating partners and have more babies. Eventually, there are more long-necked giraffes in the population.

28. Which parts of an atom are located inside the nucleus?
(Average Rigor) (Skill 7.1)

A.   electrons and neutrons.
B.   protons and neutrons.
C.   protons only.
D.   neutrons only.

29. What cell organelle contains the cell's stored food?
(Average Rigor) (Skill 7.1)

A.   Vacuoles.
B.   Golgi Apparatus.
C.   Ribosomes.
D.   Lysosomes.

30. Which of the following is not a nucleotide?
(Rigorous) (Skill 7.2)

A.   adenine.
B.   alanine.
C.   cytosine.
D.   guanine.

31. Identify the correct sequence of the organization of living things from lower to higher order:
(Rigorous) (Skill 7.3)

A.   Cell, Organelle, Organ, Tissue, System, Organism.
B.   Cell, Tissue, Organ, Organelle, System, Organism.
C.   Organelle, Cell, Tissue, Organ, System, Organism.
D.   Organelle, Tissue, Cell, Organ, System, Organism.

32. A duck's webbed feet are examples of_____.
(Rigorous) (Skill 7.6)

A.   mimicry.
B.   structural adaptation.
C.   protective resemblance.
D.   protective coloration.

33. A product of anaerobic respiration in animals is

_____.
(Easy) (Skill 7.6)

A.   carbon dioxide.
B.   lactic acid.
C.   oxygen.
D.   sodium chloride

34. Which change does *not* affect enzyme rate?
(Rigorous) (Skill 7.6)

A.   Increase the temperature.
B.   Add more substrate.
C.   Adjust the pH.
D.   Use a larger cell.

35. The first stage of mitosis is called _____ .
(Average Rigor) (Skill 8.3)

A. telophase.
B. anaphase.
C. prophase.
D. mitophase.

36. Which process(es) result(s) in a haploid chromosome number?
(Rigorous) (Skill 8.3)

A.   Mitosis.
B.   Meiosis.
C.   Both mitosis and meiosis.
D.   Neither mitosis nor meiosis.

37. A series of experiments on pea plants formed by _____ showed that two invisible markers existed for each trait, and one marker dominated the other.
(Rigorous) (Skill 8.4)

A.   Pasteur.
B.   Watson and Crick.
C.   Mendel.
D.   Mendeleev.

38. A white flower is crossed with a red flower. Which of the following is a sign of incomplete dominance?
(Average Rigor) (Skill 8.4)

A.   Pink flowers.
B.   Red flowers.
C.   White flowers.
D.   No flowers.

39. A wrasse (fish) cleans the teeth of other fish by eating away plaque. This is an example of _____ between the fish.
(Easy) (Skill 9.3)

A.   parasitism.
B.   symbiosis (mutualism).
C.   competition.
D.   predation.

40. Which of the following animals would be most likely to live in a tropical rain forest?
(Easy) (Skill 9.6)

A.   Reindeer.
B.   Monkeys.
C.   Puffins.
D.   Bears.

41. Which of the following is the longest (largest) unit of geological time?
(Easy) (Skill 10.4)

A.   Solar Year.
B.   Epoch.
C.   Period.
D.   Era.

42. The chemical equation for water formation is: $2H_2 + O_2 \rightarrow 2H_2O$. Which of the following is an *incorrect* interpretation of this equation?
(Rigorous) (Skill 11.5)

A.   Two moles of hydrogen gas and one mole of oxygen gas combine to make two moles of water.
B.   Two grams of hydrogen gas and one gram of oxygen gas combine to make two grams of water.
C.   Two molecules of hydrogen gas and one molecule of oxygen gas combine to make two molecules of water.
D.   Four atoms of hydrogen (combined as a diatomic gas) and two atoms of oxygen (combined as a diatomic gas) combine to make two molecules of water.

43. Which of the following will not change in a chemical reaction?
(Average Rigor) (11.1)

A. Number of moles of products.
B. Atomic number of one of the reactants.
C. Mass (in grams) of one of the reactants.
D. Rate of reaction.

44.Carbon bonds with hydrogen by _____ .
(Rigorous) (Skill 11.4)

A. ionic bonding.
B. non-polar covalent bonding.
C. polar covalent bonding.
D. strong nuclear force.

45. Which of the following is found in the least abundance in organic molecules?
(Average Rigor) (Skill 11.4)

A. Phosphorus.
B. Potassium.
C. Carbon.
D. Oxygen.

46. The elements in the modern Periodic Table are arranged _____ .
(Average Rigor) (Skill 11.5)

A. in numerical order by atomic number.
B. randomly.
C. in alphabetical order by chemical symbol.
D. in numerical order by atomic mass.

47. Which of the following is *not* a property of metalloids?
(Rigorous) (Skill 11.5)

A. Metalloids are solids at standard temperature and pressure.
B. Metalloids can conduct electricity to a limited extent.
C. Metalloids are found in groups 13 through 17.
D. Metalloids all favor ionic bonding.

48. The measure of the pull of the earth's gravity on an object is called _____.
(Average Rigor) (Skill 12.1)

A. mass number.
B. atomic number.
C. mass.
D. weight.

49. The Law of Conservation of Energy states that _____.
(Rigorous) (Skill 12.1)

A. There must be the same number of products and reactants in any chemical equation.
B. Objects always fall toward large masses such as planets.
C. Energy is neither created nor destroyed, but may change form.
D. Lights must be turned off when not in use, by state regulation.

50. What are the most significant and prevalent elements in the biosphere?
(Rigorous) (Skill 12.1)

A. Carbon, Hydrogen, Oxygen, Nitrogen, Phosphorus.
B. Carbon, Hydrogen, Sodium, Iron, Calcium.
C. Carbon, Oxygen, Sulfur, Manganese, Iron.
D. Carbon, Hydrogen, Oxygen, Nickel, Sodium, Nitrogen.

51. Which of the following is a correct definition for 'chemical equilibrium'?
(Rigorous) (Skill 12.1)

A. Chemical equilibrium is when the forward and backward reaction rates are equal. The reaction may continue to proceed forward and backward.
B. Chemical equilibrium is when the forward and backward reaction rates are equal, and equal to zero. The reaction does not continue.
C. Chemical equilibrium is when there are equal quantities of reactants and products.
D. Chemical equilibrium is when acids and bases neutralize each other fully.

52. What is specific gravity?
(Easy) (Skill 12.1)

A. The mass of an object.
B. The ratio of the density of a substance to the density of water.
C. Density.
D. The pull of the earth's gravity on an object.

53. Vinegar is an example of a _____ .
(Average Rigor) (Skill 12.4)

A. strong acid.
B. strong base.
C. weak acid.
D. weak base.

54. Catalysts assist reactions by _____ .
(Average Rigor) (Skill 12.5)

A. lowering effective activation energy.
B. maintaining precise pH levels.
C. keeping systems at equilibrium.
D. adjusting reaction speed.

55. Energy is measured with the same units as _____ .
(Easy) (Skill 13.1)

A. force.
B. momentum.
C. work.
D. power.

56. A ball rolls down a smooth hill. You may ignore air resistance. Which of the following is a true statement?
(Rigorous) (Skill 13.2)

A. The ball has more energy at the start of its descent than just before it hits the bottom of the hill, because it is higher up at the beginning.
B. The ball has less energy at the start of its descent than just before it hits the bottom of the hill, because it is moving more quickly at the end.
C. The ball has the same energy throughout its descent, because positional energy is converted to energy of motion.
D. The ball has the same energy throughout its descent, because a single object (such as a ball) cannot gain or lose energy.

57. A long silver bar has a temperature of 50 degrees Celsius at one end and 0 degrees Celsius at the other end. The bar will reach thermal equilibrium (barring outside influence) by the process of heat _____.
(Rigorous) (Skill 13.2)

A. conduction.
B. convection.
C. radiation.
D. phase change.

58. If the volume of a confined gas is increased, what happens to the pressure of the gas? You may assume that the gas behaves ideally, and that temperature and number of gas molecules remain constant.
(Easy) (Skill 13.4)

A. The pressure increases.
B. The pressure decreases.
C. The pressure stays the same.
D. There is not enough information given to answer this question.

59. A Newton is fundamentally a measure of _____ .
(Average Rigor) (Skill 13.5)

A. force.
B. momentum.
C. energy.
D. gravity.

60. Newton's Laws are taught in science classes because

_____.
(Average Rigor) (Skill 14.1)

A. they are the correct analysis of inertia, gravity, and forces.
B. they are a close approximation to correct physics, for usual Earth conditions.
C. they accurately incorporate relativity into studies of forces.
D. Newton was a well-respected scientist in his time.

61. Which of the following is most accurate?
(Rigorous) (Skill 14.1)

A. Mass is always constant; weight may vary by location.
B. Mass and weight are both always constant.
C. Weight is always constant; mass may vary by location.
D. Mass and weight may both vary by location.

62. All of the following are considered Newton's Laws except for:
(Average Rigor) (Skill 14.1)

A. An object in motion will continue in motion unless acted upon by an outside force.
B. For every action force, there is an equal and opposite reaction force.
C. Nature abhors a vacuum.
D. Mass can be considered the ratio of force to acceleration.

63. All of the following measure energy except for _____
(Rigorous) (Skill 14.5)

A. joules.
B. calories.
C. watts.
D. ergs.

64. Sound can be transmitted in all of the following except _____.
(Easy) (Skill 15.1)

A. air.
B. water.
C. a diamond.
D. a vacuum.

65. Sound waves are produced by _____.
(Easy) (Skill 15.2)

A. pitch.
B. noise.
C. vibrations.
D. sonar.

66. The speed of light is different in different materials. This is a result of _____.
(Average Rigor) (Skill 15.3)

A. interference.
B. refraction.
C. reflection.
D. relativity.

67. A converging lens produces a real image _____.
(Rigorous) (Skill 15.4)

A. always.
B. never.
C. when the object is within one focal length of the lens.
D. when the object is further than one focal length from the lens.

68. The Doppler Effect is associated most closely with which property of waves?
(Rigorous) (Skill 15.5)

A.   amplitude.
B.   wavelength.
C.   frequency.
D.   intensity.

69. Which of the following is *not* a necessary characteristic of living things?
(Average Rigor) (Skill 16.1)

A.   Movement.
B.   Reduction of local entropy.
C.   Ability to cause change in local energy form.
D.   Reproduction.

70. Resistance is measured in units called _____ .
(Easy) (Skill 16.2)

A.   watts.
B.   volts.
C.   ohms.
D.   current.

71. Which is the correct order of methodology?
(Average Rigor) (Skill 17.1)

**1. collecting data
2. planning a controlled experiment
3. drawing a conclusion
4. hypothesizing a result
5. re-visiting a hypothesis to answer a question**

A.   1,2,3,4,5
B.   4,2,1,3,5
C.   4,5,1,3,2
D.   1,3,4,5,2

72. Which of the following is *not* considered ethical behavior for a scientist?
(Average Rigor) (Skill 17.1)

A.   Using unpublished data and citing the source.
B.   Publishing data before other scientists have had a chance to replicate results.
C.   Collaborating with other scientists from different laboratories.
D.   Publishing work with an incomplete list of citations.

73. Extensive use of antibacterial soap has been found to increase the virulence of certain infections in hospitals. Which of the following might be an explanation for this phenomenon?
(Average Rigor) (Skill 17.1)

A.   Antibacterial soaps do not kill viruses.
B.   Antibacterial soaps do not incorporate the same antibiotics used as medicine.
C.   Antibacterial soaps kill a lot of bacteria, and only the hardiest ones survive to reproduce.
D.   Antibacterial soaps can be very drying to the skin.

74. When is a hypothesis formed?
(Easy) (Skill 17.2)

A.   Before the data is taken.
B.   After the data is taken.
C.   After the data is analyzed.
D.   Concurrent with graphing the data.

75. Which of the following is *not* an acceptable way for a student to acknowledge sources in a laboratory report?
(Easy) (Skill 17.2)

A.   The student tells his/her teacher what sources s/he used to write the report.

B.   The student uses footnotes in the text, with sources cited, but not in correct MLA format.

C.   The student uses endnotes in the text, with sources cited, in correct MLA format.

D.   The student attaches a separate bibliography, noting each use of sources.

76. In a laboratory report, what is the abstract?
(Easy) (Skill 17.4)

A.   The abstract is a summary of the report, and is the first section of the report.

B.   The abstract is a summary of the report, and is the last section of the report.

C.   The abstract is predictions for future experiments, and is the first section of the report.

D.   The abstract is predictions for future experiments, and is the last section of the report.

77. Identify the control in the following experiment: A student had four corn plants and was measuring photosynthetic rate (by measuring growth mass). Half of the plants were exposed to full (constant) sunlight, and the other half were kept in 50% (constant) sunlight.
(Average Rigor) (Skill 17.4)

A.   The control is a set of plants grown in full (constant) sunlight.

B.   The control is a set of plants grown in 50% (constant) sunlight.

C.   The control is a set of plants grown in the dark.

D.   The control is a set of plants grown in a mixture of natural levels of sunlight.

78. In an experiment measuring the growth of bacteria at different temperatures, what is the independent variable?
(Rigorous) (Skill 17.4)

A.   Number of bacteria.
B.   Growth rate of bacteria.
C.   Temperature.
D.   Light intensity.

79. A scientific law_____.
(Easy) (Skill 17.4)

A.   proves scientific accuracy.
B.   may never be broken.
C.   may be revised in light of new data.
D.   is the result of one excellent experiment.

80. Amino acids are carried to the ribosome in protein synthesis   by _____ .
(Rigorous) (Skill 18.1)

A.   transfer RNA (tRNA).
B.   messenger RNA (mRNA).
C.   ribosomal RNA (rRNA).
D.   transformation RNA (trRNA).

81. Which is the most desirable tool to use to heat substances in a middle school laboratory?
(Average Rigor) (Skill 18.1)

A.   Alcohol burner.
B.   Freestanding gas burner.
C.   Bunsen burner.
D.   Hot plate.

82. Chemicals should be stored
(Easy) (Skill 18.1)

A.   in the principal's office.
B.   in a dark room.
C.   in an off-site research facility.
D.   according to their reactivity with other substances.

83. When measuring the volume of water in a graduated cylinder, where does one read the measurement?
(Average Rigor) (Skill 18.2)

A.   At the highest point of the liquid.
B.   At the bottom of the meniscus curve.
C.   At the closest mark to the top of the liquid.
D.   At the top of the plastic safety ring.

84. Who should be notified in the case of a serious chemical spill?
(Easy) (Skill 18.2)

A.   The custodian.
B.   The fire department or their municipal authority.
C.   The science department chair.
D.   The School Board.

85. In a science experiment, a student needs to dispense very small measured amounts of liquid into a well-mixed solution. Which of the following is the best choice for his/her equipment to use?
(Average Rigor) (Skill 18.2)

A.   Buret with Buret Stand, Stir-plate, Stirring Rod, Beaker.
B.   Buret with Buret Stand, Stir-plate, Beaker.
C.   Volumetric Flask, Dropper, Graduated Cylinder, Stirring Rod.
D.   Beaker, Graduated Cylinder, Stir-plate.

86. A laboratory balance is most appropriately used to measure the mass of which of the following?
(Easy) (Skill 18.2)

A.   Seven paper clips.
B.   Three oranges.
C.   Two hundred cells.
D.   One student's elbow.

87. Who determines the laws regarding the use of safety glasses in the classroom?
(Average Rigor) (Skill 18.4)

A.   The state.
B.   The school site.
C.   The Federal government.
D.   The district level.

88. Formaldehyde should not be used in school laboratories for the following reason:
(Easy) (Skill 18.4)

A.   it smells unpleasant.
B.   it is a known carcinogen.
C.   it is expensive to obtain.
D.   it is explosive.

89. Experiments may be done with any of the following animals except _____ .
(Easy) (Skill 18.4)

A.   birds.
B.   invertebrates.
C.   lower order life.
D.   frogs.

90. Which of the following is the worst choice for a school laboratory activity?
(Average Rigor) (Skill 18.4)

A.   A genetics experiment tracking the fur color of mice.
B.   Dissection of a preserved fetal pig.
C.   Measurement of goldfish respiration rate at different temperatures.
D.   Pithing a frog to watch the circulatory system.

91. In which situation would a science teacher be legally liable?
(Rigorous) (Skill 18.5)

A.   The teacher leaves the classroom for a telephone call and a student slips and injures him/herself.
B.   A student removes his/her goggles and gets acid in his/her eye.
C.   A faulty gas line in the classroom causes a fire.
D.   A student cuts him/herself with a dissection scalpel.

92. Which of these is the best example of 'negligence'?
(Average Rigor) (Skill 18.5)

A.   A teacher fails to give oral instructions to those with reading disabilities.
B.   A teacher fails to exercise ordinary care to ensure safety in the classroom.
C.   A teacher displays inability to supervise a large group of students.
D.   A teacher reasonably anticipates that an event may occur, and plans accordingly.

93. Which item should always be used when handling glassware?
(Easy) (Skill 18.5)

A.   Tongs.
B.   Safety goggles.
C.   Gloves.
D.   Buret stand.

94. Accepted procedures for preparing solutions involve the use of _____ .
(Average Rigor) (Skill 18.5)

A.   alcohol.
B.   hydrochloric acid.
C.   distilled water.
D.   tap water.

95. When designing a scientific experiment, a student considers all the factors that may influence the results. The process goal is to _____.
(Easy) (Skill 19.2)

A.   recognize and manipulate independent variables.
B.   recognize and record independent variables.
C.   recognize and manipulate dependent variables.
D.   recognize and record dependent variables.

96. What is the scientific method?
(Easy) (Skill 19.3)

A.   It is the process of doing an experiment and writing a laboratory report.
B.   It is the process of using open inquiry and repeatable results to establish theories.
C.   It is the process of reinforcing scientific principles by confirming results.
D.   It is the process of recording data and observations.

97. Which of the following data sets is properly represented by a bar graph?
(Average Rigor) (Skill 19.3)

A.   Number of people choosing to buy cars, vs. Color of car bought.
B.   Number of people choosing to buy cars, vs. Age of car customer.
C.   Number of people choosing to buy cars, vs. Distance from car lot to customer home.
D.   Number of people choosing to buy cars, vs. Time since last car purchase.

98. For her first project of the year, a student is designing a science experiment to test the effects of light and water on plant growth. You should recommend that she _____.
(Average Rigor) (Skill 19.4)

A.   manipulate the temperature also.
B.   manipulate the water pH also.
C.   determine the relationship between light and water unrelated to plant growth.
D.   omit either water or light as a variable.

99. The theory of 'continental drift' is supported by which of the following?
(Rigorous) (Skill 20.3)

A.   The way the shapes of South America and Europe fit together.
B.   The way the shapes of Europe and Asia fit together.
C.   The way the shapes of South America and Africa fit together.
D.   The way the shapes of North America and Antarctica fit together.

Answer Key

1. B	45. B	89. A
2. B	46. A	90. D
3. B	47. D	91. A
4. A	48. D	92. B
5. B	49. C	93. B
6. B	50. A	94. C
7. C	51. A	95. A
8. D	52. B	96. B
9. C	53. C	97. A
10. D	54. A	98. D
11. B	55. C	99. C
12. C	56. C	
13. D	57. A	
14. D	58. B	
15. B	59. A	
16. D	60. B	
17. A	61. A	
18. C	62. C	
19. B	63. C	
20. B	64. D	
21. D	65. C	
22. A	66. B	
23. A	67. D	
24. B	68. C	
25. B	69. A	
26. C	70. C	
27. B	71. B	
28. B	72. D	
29. A	73. C	
30. B	74. A	
31. C	75. A	
32. B	76. A	
33. B	77. A	
34. D	78. C	
35. C	79. C	
36. B	80. A	
37. C	81. D	
38. A	82. D	
39. B	83. B	
40. B	84. B	
41. D	85. B	
42. B	86. A	
43. B	87. A	
44. C	88. B	

Rigor Table

Easy Rigor 20 %	Average Rigor 40%	Rigorous 40%
7, 8, 10, 17, 20, 33, 39, 40, 41, 55, 58, 64, 65, 70, 74, 75, 76, 79, 82, 84, 86, 88, 89, 93, 95, 96	2, 3, 5, 13, 16, 19, 22, 26, 27, 28, 29, 35, 38, 43, 45, 46, 48, 53, 54, 59, 60, 62, 66, 69, 71, 72, 73, 77, 81, 83, 85, 87, 90, 92, 94, 97, 98	1, 4, 6, 9, 11, 12, 14, 15, 18, 21, 23, 24, 25, 30, 31, 32, 34, 36, 37, 44, 47, 49, 50, 51, 56, 57, 61, 63, 67, 68, 78, 80, 91, 99

Rationales with Sample Questions

1. What is the main difference between the 'condensation hypothesis' and the 'tidal hypothesis' for the origin of the solar system?
(Rigorous) (Skill 1.1)

A. The tidal hypothesis can be tested, but the condensation hypothesis cannot.

B. The tidal hypothesis proposes a near collision of two stars pulling on each other, but the condensation hypothesis proposes condensation of rotating clouds of dust and gas.

C. The tidal hypothesis explains how tides began on planets such as Earth, but the condensation hypothesis explains how water vapor became liquid on Earth.

D. The tidal hypothesis is based on Aristotelian physics, but the condensation hypothesis is based on Newtonian mechanics.

B. The tidal hypothesis proposes a near collision of two stars pulling on each other, but the condensation hypothesis proposes   condensation of rotating clouds of dust and gas.

Most scientists believe the 'condensation hypothesis,' i.e. that the solar system began when rotating clouds of dust and gas condensed into the sun and planets. A minority opinion is the 'tidal hypothesis,' i.e. that the sun almost collided with a large star. Because both of these hypotheses deal with ancient, unrepeatable events, neither can be tested, eliminating answer (A). Note that both 'tidal' and 'condensation' have additional meanings in science, but those are not relevant here, eliminating answer (C). Both hypotheses are based on best guesses using modern physics, eliminating answer (D). Therefore, the **answer is (B).**

2. Which of the following is the best definition for 'meteorite'?
(Average Rigor) (Skill 1.2)

A. A meteorite is a mineral composed of mica and feldspar.

B. A meteorite is material from outer space that has struck the earth's surface.

C. A meteorite is an element that has properties of both metals and nonmetals.

D. A meteorite is a very small unit of length measurement.

B. A meteorite is material from outer space, that has struck the earth's surface.

Meteoroids are pieces of matter in space, composed of particles of rock and metal. If a meteoroid travels through the earth's atmosphere, friction causes burning and a "shooting star"—i.e. a meteor. If the meteor strikes the earth's surface, it is known as a meterorite. Note that although the suffix –ite often means a mineral, answer (A) is incorrect. Answer (C) refers to a 'metalloid' rather than a 'meteorite', and answer (D) is simply a misleading pun on 'meter'. Therefore, the **answer is (B)**.

3. The force of gravity on earth causes all bodies in free fall to _____
(Average Rigor) (Skill 1.3)

A. fall at the same speed.

B. accelerate at the same rate.

C. reach the same terminal velocity.

D. move in the same direction.

B. Accelerate at the same rate.

Gravity causes approximately the same acceleration on all falling bodies close to earth's surface.
More massive
bodies continue to accelerate at this rate for longer, before their air resistance is great enough to cause terminal velocity, so answers (A) and (C) are eliminated. Bodies on different parts of the planet move in different directions (always toward the center of mass of earth), so answer (D) is eliminated. Thus, the **answer is (B)**.

4. The electromagnetic radiation with the longest wave length is/are
_____
(Rigorous) (Skill 1.6)

A. radio waves.

B. red light.

C. X-rays.

D. ultraviolet light.

A. Radio waves.

Radio waves have longer wave lengths (and smaller frequencies) than visible light, which in turn has longer wave lengths than ultraviolet or X-ray radiation. Radio waves are considered much less harmful (less energetic, i.e. lower frequency) than ultraviolet or X-ray radiation. The correct answer is **therefore (A)**.

5. When heat is added to most solids, they expand. Why is this the case?
(Average Rigor) (Skill 2.1)

A. The molecules get bigger.

B. The faster molecular motion leads to greater distance between the molecules.

C. The molecules develop greater repelling electric forces.

D. The molecules form a more rigid structure.

B. The faster molecular motion leads to greater distance between the molecules.

The atomic theory of matter states that matter is made up of tiny, rapidly moving particles. Temperature is a measure of average kinetic energy of the particles. Warmer molecules move more rapidly, separating farther from each other. The individual molecules do not get bigger, by conservation of mass, eliminating answer (A). The molecules do not develop greater repelling electric forces, eliminating answer (C). Occasionally, molecules form a more rigid structure when becoming colder and freezing (such as water)—but this gives rise to the exceptions to heat expansion, so it is not relevant here, eliminating answer (D). Therefore, the **answer is (B)**.

6. Which of the following is *not* true about phase change in matter?
(Rigorous) (Skill 2.1)

A. Solid water and liquid ice can coexist at water's freezing point.

B. At 7 degrees Celsius, water is always in liquid phase.

C. Matter changes phase when enough energy is gained or lost.

D. Different phases of matter are characterized by differences in molecular motion.

B. At 7 degrees Celsius, water is always in liquid phase.

According to the molecular theory of matter, molecular motion determines the 'phase' of the matter, and the energy in the matter determines the speed of molecular motion. Solids have vibrating molecules that are in fixed relative positions; liquids have faster molecular motion than their solid forms, and the molecules may move more freely but must still be in contact with one another; gases have even more energy and more molecular motion. At the 'freezing point' or 'boiling point' of a substance, both relevant phases may be present. Pressure changes, in addition to temperature changes, can cause phase changes. For example, nitrogen can be liquefied under high pressure, even though its boiling temperature is very low. Therefore, the **correct answer must be (B)**. Water may be a liquid at that temperature, but it may also be a solid, depending on ambient pressure.

7. What is the most accurate description of the Water Cycle?
(Easy) (Skill 2.2)

A. Rain comes from clouds, filling the ocean. The water then evaporates and becomes clouds again.

B. Water circulates from rivers into groundwater and back, while water vapor circulates in the atmosphere.

C. Water is conserved except for chemical or nuclear reactions, and any drop of water could circulate through clouds, rain, ground-water, and surface-water.

D. Weather systems cause chemical reactions to break water into its atoms.

C. Water is conserved except for chemical or nuclear reactions, and any drop of water could circulate through clouds, rain, ground-water, and surface-water.

All natural chemical cycles, including the Water Cycle, depend on the principle of Conservation of Mass. Any drop of water may circulate through the hydrologic system, ending up in a cloud, as rain, or as surface- or ground-water. Although answers (A) and (B) describe parts of the water cycle, the most comprehensive and correct **answer is (C)**.

8. What is the source for most of the United States' drinking water?
(Easy) (Skill 2.2)

A. Desalinated ocean water.

B. Surface water (lakes, streams, mountain runoff).

C. Rainfall into municipal reservoirs.

D. Groundwater.

D. Groundwater.

Groundwater currently provides drinking water for 53% of the population of the United States. The other answer choices can be used for drinking water, but they are not the most widely used. Therefore, **the answer is (D)**.

9. The salinity of ocean water is closest to _____ .
(Rigorous) (Skill 2.5)

A. 0.035 %

B. 0.35 %

C. 3.5 %

D. 35 %

C. 3.5 %

Salinity, or concentration of dissolved salt, can be measured in mass ratio (i.e. mass of salt divided by mass of sea water). For Earth's oceans, the salinity is approximately 3.5 %, or 35 parts per thousand. Note that answers (A) and (D) can be eliminated, because (A) is so dilute as to be hardly saline, while (D) is so concentrated that it would not support ocean life. Therefore, the **answer is (C)**.

10. The transfer of heat by electromagnetic waves is called _____
(Easy) (Skill 3.2)

A. conduction.

B. convection.

C. phase change.

D. radiation.

D. Radiation.

Heat transfer via electromagnetic waves (which can occur even in a vacuum) is called radiation. (Heat can also be transferred by direct contact (conduction), by fluid current (convection), and by matter changing phase, but these are not relevant here.) The answer to this question is **therefore (D)**.

11. Which of the following instruments measures wind speed?
(Rigorous) (Skill 13.7)

A. A barometer.

B. An anemometer.

C. A wind sock

D. A weather vane.

B. An anemometer.

An anemometer is a device to measure wind speed, while a barometer measures pressure and both a wind sock wind and a weather vane indicate wind direction. This is consistent only with **answer (B).**

12. A cup of hot liquid and a cup of cold liquid are both sitting in a room at comfortable room temperature and humidity. Both cups are thin plastic. Which of the following is a true statement?
(Rigorous) (Skill 3.7)

A. There will be fog on the outside of the hot liquid cup, and also fog on the outside of the cold liquid cup.

B. There will be fog on the outside of the hot liquid cup, but not on the cold liquid cup.

C. There will be fog on the outside of the cold liquid cup, but not on the hot liquid cup.

D. There will not be fog on the outside of either cup.

C. There will be fog on the outside of the cold liquid cup, but not on the hot liquid cup.

Fog forms on the outside of a cup when the contents of the cup are colder than the surrounding air, and the cup material is not a perfect insulator. This happens because the air surrounding the cup is cooled to a lower temperature than the ambient room, so it has a lower saturation point for water vapor. Although the humidity had been reasonable in the warmer air, when that air circulates near the colder region and cools, water condenses onto the cup's outside surface. Therefore, the correct **answer is (C).**

13. Which of the following types of rock is made from magma?
(Average Rigor) (Skill 4.1)

A. Fossils.

B. Sedimentary.

C. Metamorphic.

D. Igneous.

D. Igneous.

Igneous rocks are formed from magma and magma is so hot that any organisms trapped by it are destroyed. Metamorphic rocks are formed by high temperatures and great pressures. When fluid sediments are transformed into solid sedimentary rocks, the process is known as lithification. Fossils are found only in sedimentary rocks.
The **answer is (D).**

14. Igneous rocks can be classified according to which of the following?
(Rigorous) (Skill 4.1)

A. Texture.

B. Composition.

C. Formation process.

D. All of the above.

D. All of the above.

Igneous rocks, which form from the crystallization of molten lava, are classified according to many of their characteristics, including texture, composition, and how they were formed. Therefore, **the answer is (D).**

15. Lithification refers to the process by which unconsolidated sediments are transformed into _____
(Rigorous) (Skill 4.2)

A. metamorphic rocks.

B. sedimentary rocks.

C. igneous rocks.

D. lithium oxide.

B. Sedimentary rocks.

Lithification is the process of sediments coming together to form rocks, i.e. sedimentary rock formation. Metamorphic and igneous rocks are formed via other processes (heat and pressure or volcano, respectively).
Therefore, the **answer must be (B)**.

16. Which of these is a true statement about loamy soil?
(Average Rigor) (Skill 4.3)

A. Loamy soil is gritty and porous.

B. Loamy soil is smooth and a good barrier to water.

C. Loamy soil is hostile to microorganisms.

D. Loamy soil is velvety and clumpy.

D. Loamy soil is velvety and clumpy.

The three classes of soil by texture are: Sandy (gritty and porous), Clay (smooth, greasy, and most impervious to water), and Loamy (velvety, clumpy, and able to hold water and let water flow through). In addition, loamy soils are often the most fertile soils. Therefore, the **answer must be (D)**.

17. _____ are cracks in the plates of the earth's crust, along which the plates move.
(Easy) (Skill 4.5)

A. Faults

B. Ridges

C. Earthquakes

D. Volcanoes

A. Faults.

Faults are cracks in the earth's crust, and when the earth moves, an earthquake results. The answer to this question must **therefore be (A).**

18. Which of the following is *not* a type of volcano?
(Rigorous) (Skill 4.5)

A. Shield Volcanoes.

B. Composite Volcanoes.

C. Stratus Volcanoes.

D. Cinder Cone Volcanoes.

C. Stratus Volcanoes.

There are three types of volcanoes. Shield volcanoes (A) are associated with non-violent eruptions and repeated lava flow over time. Composite volcanoes (B) are built from both lava flow and layers of ash and cinders. Cinder cone volcanoes (D) are associated with violent eruptions. **'Stratus' (C)** is a type of cloud, so it is the correct answer to this question.

19. Fossils are usually found in _____ rock.
(Average Rigor) (Skill 4.6)

A. igneous.

B. sedimentary.

C. metamorphic.

D. cumulus.

B. Sedimentary

Fossils are formed by layers of dirt and sand settling around organisms, hardening, and taking an imprint of the dead organisms. Rocks that form from layers of settling dirt and sand are sedimentary rock. Igneous rock is formed from molten lava, while metamorphic rock is formed from any rock under very high temperature and pressure. 'Cumulus' is a descriptor for clouds. The best answer is **therefore (B)**.

20. Which of the following is the most accurate definition of a nonrenewable resource?
(Easy) (Skill 5.1)

A. A nonrenewable resource is never replaced once used.

B. A nonrenewable resource is replaced on a timescale that is very long relative to human life-spans.

C. A nonrenewable resource is a resource that can only be manufactured by humans.

D. A nonrenewable resource is a species that has already become extinct.

B. A nonrenewable resource is replaced on a timescale that is very long relative to human life-spans.

Renewable resources are those that are renewed, or replaced, in time for the same or the next generation of humans to use more of them. Examples include fast-growing plants, animals, or oxygen gas. Nonrenewable resources are those that renew themselves only on very long timescales, usually geologic timescales. Examples include minerals, metals, or fossil fuels. Therefore, the **correct answer is (B)**.

21. Which of the following is *not* a common type of acid in 'acid rain' or acidified surface water?
(Rigorous) (Skill 5.4)

A. Nitric acid.

B. Sulfuric acid.

C. Carbonic acid.

D. Hydrofluoric acid.

D. Hydrofluoric acid.

Acid rain forms predominantly from pollutant oxides in the air (usually nitrogen-based NO_x or sulfur-based SO_x), which become hydrated into their acids (nitric or sulfuric acid). Because of increased levels of carbon dioxide pollution, carbonic acid is also common in acidified surface water environments. Hydrofluoric acid can be found, but it is much less common. Therefore, the **answer is (D)**.

22. Laboratory researchers have classified fungi as distinct from plants because the cell walls of fungi
(Average Rigor) (Skill 6.1)

A. contain chitin.

B. contain yeast.

C. are more solid.

D. are less solid.

A. Contain chitin.

Kingdom Fungi consists of organisms that are eukaryotic, multicellular, absorptive consumers. They have a chitin cell wall, which is the only universally present feature in fungi that is never present in plants. Thus, the **answer is (A)**.

23. Which kingdom is comprised of organisms made of one cell with no nuclear membrane?
(Rigorous) (Skill 6.1)

A. Monera.

B. Protista.

C. Fungi.

D. Algae.

A. Monera.

Algae are not a kingdom of their own. Some algae are in monera, the kingdom that consists of unicellular prokaryotes with no true nucleus. Protista and fungi are both eukaryotic, with true nuclei, and are sometimes multi-cellular. Therefore, the **answer is (A)**.

24. Animals with a notochord or a backbone are in the phylum
(Rigorous) (Skill 6.1)

A. arthropoda.

B. chordata.

C. mollusca.

D. mammalia.

B. Chordata.

The phylum arthropoda contains spiders and insects and phylum mollusca contain snails and squid. Both of these phyla have exoskeletons rather than backbones. Mammalia is a class in the phylum chordata. The **answer is (B).**

25. The scientific name *Canis familiaris* refers to the animal's
_____.
(Rigorous) (Skill 6.3)

A. kingdom and phylum.

B. genus and species.

C. class and species.

D. type and family.

B. Genus and species.

Genus and species are the most specific way to identify an organism. Usually the genus is capitalized and the species, immediately following, is not. 'Canis' is the genus for dogs, or canines. Therefore, the **answer must be (B)**. Knowing there is no such kingdom as 'Canis,' and no category 'type' in official taxonomy could eliminate answers (A) and (D).

26. Members of the same animal species _____
(Average Rigor) (Skill 6.3)

A. look identical.

B. never adapt differently.

C. are able to reproduce with one another.

D. are found in the same location.

C. Are able to reproduce with one another.

Although members of the same animal species may look alike (A), adapt alike (B), or be found near one another (D), the only requirement to be part of the same species is that they be able to reproduce with one another. Therefore, the **answer is (C)**.

27. Which of the following is a correct explanation for scientific 'evolution'?
(Average Rigor) (Skill 6.5)

A. Giraffes need to reach higher for leaves to eat, so their necks stretch. The giraffe babies are then born with longer necks. Eventually, there are more long-necked giraffes in the population.

B. Giraffes with longer necks are able to reach more leaves, so they eat more and have more babies than other giraffes. Eventually, there are more long-necked giraffes in the population.

C. Giraffes want to reach higher for leaves to eat, so they release enzymes into their bloodstream, which in turn causes fetal development of longer-necked giraffes. Eventually, there are more long-necked giraffes in the population.

D. Giraffes with long necks are more attractive to other giraffes, so they get the best mating partners and have more babies. Eventually, there are more long-necked giraffes in the population.

B. Giraffes with longer necks are able to reach more leaves, so they eat more and have more babies than other giraffes. Eventually, there are more long-necked giraffes in the population.

Evolution occurs via natural selection. Organisms with a life/reproductive advantage will produce more offspring. Over many generations, this changes the proportions of the population. It is impossible for a stretched neck (A) or a fervent desire (C) to result in a biologically mutated baby. Although there are traits that are naturally selected because of mate attractiveness and fitness (D), this is not the primary situation here, **so answer (B) is the best choice.**

28. Which parts of an atom are located inside the nucleus?
(Average Rigor) (Skill 7.1)

A. electrons and neutrons.

B. protons and neutrons.

C. protons only.

D. neutrons only.

B. Protons and Neutrons.

Protons and neutrons are located in the nucleus, while electrons move around outside the nucleus. This is consistent only with **answer (B)**.

29. What cell organelle contains the cell's stored food?
(Average Rigor) (Skill 7.1)

A. Vacuoles.

B. Golgi Apparatus.

C. Ribosomes.

D. Lysosomes.

A. Vacuoles.

The vacuoles hold stored food (and water and pigments). The Golgi Apparatus sorts molecules from other parts of the cell; the ribosomes are sites of protein synthesis; the lysosomes contain digestive enzymes. This is consistent only with **answer (A).**

30. Which of the following is not a nucleotide?
(Rigorous) (Skill 7.2)

A. adenine.

B. alanine.

C. cytosine.

D. guanine.

B. alanine.

Alanine is an amino acid. Adenine, cytosine, guanine, thymine, and uracil are nucleotides. The correct **answer is (B).**

31. Identify the correct sequence of the organization of living things from lower to higher order:
(Rigorous) (Skill 7.3)

A. Cell, Organelle, Organ, Tissue, System, Organism.

B. Cell, Tissue, Organ, Organelle, System, Organism.

C. Organelle, Cell, Tissue, Organ, System, Organism.

D. Organelle, Tissue, Cell, Organ, System, Organism.

C. Organelle, Cell, Tissue, Organ, System, Organism.

Organelles are parts of the cell; cells make up tissue, which makes up organs. Organs work together in systems (e.g. the respiratory system), and the organism is the living thing as a whole. Therefore, the **answer must be (C)**.

32. A duck's webbed feet are examples of
(Rigorous) (Skill 7.6)

A. mimicry.

B. structural adaptation.

C. protective resemblance.

D. protective coloration.

B. Structural adaptation.

Ducks (and other aquatic birds) have webbed feet, which makes them more efficient swimmers. Because the structure of the duck's feet adapted to its environment over generations, this is termed 'structural adaptation' which is **answer (B)**. Mimicry, protective resemblance, and protective coloration refer to other evolutionary mechanisms for survival.

33. A product of anaerobic respiration in animals is _____
(Easy) (Skill 7.6)

A. carbon dioxide.

B. lactic acid.

C. oxygen.

D. sodium chloride.

B. Lactic acid.

In animals, anaerobic respiration (i.e. respiration without the presence of oxygen) generates lactic acid as a byproduct. Oxygen is not normally a by-product of respiration, though it is a product of photosynthesis, and sodium chloride is not strictly relevant in this question. Therefore, the **answer must be (B)**.

34. Which change does *not* affect enzyme rate?
(Rigorous) (Skill 7.6)

A. Increase the temperature.

B. Add more substrate.

C. Adjust the pH.

D. Use a larger cell.

D. Use a larger cell.

Temperature, chemical amounts, and pH can all affect enzyme rate. However, the chemical reactions take place on a small enough scale that the overall cell size is not relevant. Therefore, the **answer is (D)**.

35. The first stage of mitosis is called _____
(Average Rigor) (Skill 8.3)

A. telophase.

B. anaphase.

C. prophase.

D. mitophase.

C. Prophase.

In mitosis, the division of somatic cells, prophase (**answer C**) is the stage where the cell enters mitosis. The four stages of mitosis, in order, are: prophase, metaphase, anaphase, and telophase. ("Mitophase" is not one of the steps.)

36. Which process(es) result(s) in a haploid chromosome number?
(Rigorous) (Skill 8.3)

A. Mitosis.

B. Meiosis.

C. Both mitosis and meiosis.

D. Neither mitosis nor meiosis.

B. Meiosis.

Meiosis is the division of sex cells. The resulting chromosome number is half the number of parent cells, i.e. a 'haploid chromosome number'. Mitosis, however, is the division of other cells, in which the chromosome number is the same as the parent cell chromosome number. Therefore, the **answer is (B)**.

37. A series of experiments on pea plants formed by _____ showed that two invisible markers existed for each trait, and one marker dominated the other.
(Rigorous) (Skill 8.4)

A. Pasteur.

B. Watson and Crick.

C. Mendel.

D. Mendeleev.

C. Mendel.

Gregor Mendel was a ninteenth-century Austrian botanist who derived "laws" governing inherited traits using a series of cross-breedings of pea plants. His work led to the understanding of dominant and recessive traits, carried by biological markers.

Pasteur, Watson, Crick, and Mendeleev were other scientists with different specialties. This is consistent only with **answer (C)**.

38. A white flower is crossed with a red flower. Which of the following is a sign of incomplete dominance?
(Average Rigor) (Skill 8.4)

A. Pink flowers.

B. Red flowers.

C. White flowers.

D. No flowers.

A. Pink flowers.

Incomplete dominance means that neither the red nor the white gene is strong enough to completely suppress the other. Therefore both are expressed, leading in this case to the formation of pink flowers. Therefore, the **answer is (A)**.

39. A wrasse (fish) cleans the teeth of other fish by eating away plaque. This is an example of _____ between the fish. *(Easy) (Skill 9.3)*

A. parasitism.

B. symbiosis (mutualism).

C. competition.

D. predation.

B. Symbiosis (mutualism).

When both species benefit from their interaction in their habitat, this is called 'symbiosis', or 'mutualism'. In this example, the wrasse benefits from having a source of food, and the other fish benefit by having healthier teeth. Note that 'parasitism' is when one species benefits at the expense of the other, 'competition' is when two species compete with one another for the same habitat or food, and 'predation' is when one species feeds on another. Therefore, the **answer is (B)**.

40. Which of the following animals would be most likely to live in a tropical rain forest? *(Easy) (Skill 9.6)*

A. Reindeer.

B. Monkeys.

C. Puffins.

D. Bears.

B. Monkeys.

The tropical rain forest biome is hot, humid, and very fertile—it is thought to contain almost half of the world's species. Reindeer (A), puffins (C), and bears (D), however, are usually found in much colder climates. There are several species of monkeys that thrive in hot, humid climates, so **answer (B) is correct.**

41. Which of the following is the longest (largest) unit of geological time?
(Easy) (Skill 10.4)

A. Solar Year.

B. Epoch.

C. Period.

D. Era.

D. Era.

Geological time is measured by many units, but the longest unit listed here (and indeed the longest used to describe the biological development of the planet) is the Era. Eras are subdivided into Periods, which are further divided into Epochs. Therefore, the **answer is (D).**

42. The chemical equation for water formation is: $2H_2 + O_2 \rightarrow 2H_2O$. Which of the following is an *incorrect* interpretation of this equation?
(Rigorous) (Skill 11.5)

A. Two moles of hydrogen gas and one mole of oxygen gas combine to make two moles of water.

B. Two grams of hydrogen gas and one gram of oxygen gas combine to make two grams of water.

C. Two molecules of hydrogen gas and one molecule of oxygen gas combine to make two molecules of water.

D. Four atoms of hydrogen (combined as a diatomic gas) and two atoms of oxygen (combined as a diatomic gas) combine to make two molecules of water.

B. Two grams of hydrogen gas and one gram of oxygen gas combine to make two grams of water.

In any chemical equation, the coefficients indicate the relative proportions of molecules (or atoms), or of moles of molecules. Chemicals combine in repeatable combinations of molar ratio (i.e. number of moles), but vary in mass per mole of material. Therefore, the answer must be **answer (B)**, which refers to grams, a unit of mass.

43. Which of the following will not change in a chemical reaction?
(Average Rigor) (11.1)

A. Number of moles of products.

B. Atomic number of one of the reactants.

C. Mass (in grams) of one of the reactants.

D. Rate of reaction.

B. Atomic number of one of the reactants.

Atomic number, i.e. the number of protons in a given element, is constant unless involved in a nuclear reaction. Meanwhile, the amounts (measured in moles (A) or in grams(C)) of reactants and products change over the course of a chemical reaction, and the rate of a chemical reaction (D) may change due to internal or external processes. Therefore, the **answer is (B)**.

44. Carbon bonds with hydrogen by
(Rigorous) (Skill 11.4)

A. ionic bonding.

B. non-polar covalent bonding.

C. polar covalent bonding.

D. strong nuclear force.

C. Polar covalent bonding.

Each carbon atom contains four valence electrons, while each hydrogen atom contains one valence electron. A carbon atom can bond with one or more hydrogen atoms, such that two electrons are shared in each bond. This is covalent bonding. Covalent bonds are always polar when between two non-identical atoms which is **answer (C)**. The strong nuclear force is not relevant to this problem. Ionic bonding is based on gaining or losing electrons.

45. Which of the following is found in the least abundance in organic molecules?
(Average Rigor) (Skill 11.4)

A. Phosphorus.

B. Potassium.

C. Carbon.

D. Oxygen.

B. Potassium.

Organic molecules consist mainly of carbon, hydrogen, and oxygen, with significant amounts of nitrogen, phosphorus, and often sulfur. Other elements, such as potassium, are present in much smaller quantities. Therefore, the **answer is (B)**.

46. The elements in the modern Periodic Table are arranged
(Average Rigor) (Skill 11.5)

A. in numerical order by atomic number.

B. randomly.

C. in alphabetical order by chemical symbol.

D. in numerical order by atomic mass.

A. In numerical order by atomic number.

Although the first periodic tables were arranged by atomic mass, the modern table is arranged by atomic number, i.e. the number of protons in each element which is **answer (A)**. (This allows the element list to be complete and unique.)

47. Which of the following is *not* a property of metalloids?
(Rigorous) (Skill 11.5)

A. Metalloids are solids at standard temperature and pressure.

B. Metalloids can conduct electricity to a limited extent.

C. Metalloids are found in groups 13 through 17.

D. Metalloids all favor ionic bonding.

D. Metalloids all favor ionic bonding.

Metalloids are substances that have characteristics of both metals and nonmetals, including limited conduction of electricity and being in the solid phase at standard temperature and pressure. Metalloids are found in a 'stair-step' pattern from boron in group 13 through astatine in group 17. Some metalloids, e.g. silicon, favor covalent bonding. Others, e.g. astatine, can bond ionically. Therefore, **the answer is (D).**

48. The measure of the pull of Earth's gravity on an object is called
(Average Rigor) (Skill 12.1)

A. mass number.

B. atomic number.

C. mass.

D. weight.

D. Weight.

The mass number is the total number of protons and neutrons in an atom, atomic number is the number of protons in an atom, and mass is the amount of matter in an object. The only remaining **choice is (D)**, weight, which is correct because weight is the force of gravity on an object.

49. The Law of Conservation of Energy states that
(Rigorous) (Skill 12.1)

A. there must be the same number of products and reactants in any chemical equation.

B. objects always fall toward large masses such as planets.

C. energy is neither created nor destroyed, but may change form.

D. lights must be turned off when not in use, by state regulation.

C. Energy is neither created nor destroyed, but may change form.

Answer (C) is a summary of the Law of Conservation of Energy (for non-nuclear reactions). In other words, energy can be transformed into various forms such as kinetic, potential, electric, or heat energy, but the total amount of energy remains constant. Answer (A) is untrue, as demonstrated by many synthesis and decomposition reactions. Answers (B) and (D) are not relevant in this case.

50. What are the most significant and prevalent elements in the biosphere?
(Rigorous) (Skill 12.1)

A. Carbon, Hydrogen, Oxygen, Nitrogen, Phosphorus.

B. Carbon, Hydrogen, Sodium, Iron, Calcium.

C. Carbon, Oxygen, Sulfur, Manganese, Iron.

D. Carbon, Hydrogen, Oxygen, Nickel, Sodium, Nitrogen.

A. Carbon, Hydrogen, Oxygen, Nitrogen, Phosphorus.

Organic matter (and life as we know it) is based on Carbon atoms, bonded to hydrogen and oxygen. Nitrogen and phosphorus are the next most significant elements, followed by sulfur and then trace nutrients such as iron, sodium, calcium, and others. Therefore, the **answer is (A)**.

51. Which of the following is a correct definition for 'chemical equilibrium'?
(Rigorous) (Skill 12.1)

A. Chemical equilibrium is when the forward and backward reaction rates are equal. The reaction may continue to proceed forward and backward.

B. Chemical equilibrium is when the forward and backward reaction rates are equal, and equal to zero. The reaction does not continue.

C. Chemical equilibrium is when there are equal quantities of reactants and products.

D. Chemical equilibrium is when acids and bases neutralize each other fully.

A. Chemical equilibrium is when the forward and backward reaction rates are equal. The reaction may continue to proceed forward and backward.

Chemical equilibrium is defined as when the quantities of reactants and products are at a 'steady state' and are no longer shifting, but the reaction may still proceed forward and backward. The rate of forward reaction must equal the rate of backward reaction. Note that there may or may not be equal amounts of chemicals, and that this is not restricted to a completed reaction or to an acid-base reaction. Therefore, the **answer is (A)**.

52. What is specific gravity?
(Easy) (Skill 12.1)

A. The mass of an object.

B. The ratio of the density of a substance to the density of water.

C. Density.

D. The pull of the earth's gravity on an object.

B. The ratio of the density of a substance to the density of water.

Mass is a measure of the amount of matter in an object. Density is the mass of a substance contained per unit of volume. Weight is the measure of the earth's pull of gravity on an object. The only option here is the ratio of the density of a substance to the density of water, **answer (B)**.

53. Vinegar is an example of a _____
(Average Rigor) (Skill 12.4)

A. strong acid.

B. strong base.

C. weak acid.

D. weak base.

C. Weak acid.

The main ingredient in vinegar is acetic acid, a weak acid. Vinegar is not a strong acid, such as hydrochloric acid, because it does not dissociate as fully or cause as much corrosion. It is not a base. Therefore, the **answer is (C)**.

54. Catalysts assist reactions by _____
(Average Rigor) (Skill 12.5)

A. lowering effective activation energy.

B. maintaining precise pH levels.

C. keeping systems at equilibrium.

D. adjusting reaction speed.

A. Lowering effective activation energy.

Catalysts, which are present both with reactants and with products, induce the formation of activated complexes, thereby lowering the effective activation energy. Although this often makes reactions faster, answer (D) is not as good a choice as the more generally applicable **answer (A)**, which is correct.

55. Energy is measured with the same units as _____
(Easy) (Skill 13.1)

A. force.

B. momentum.

C. work.

D. power.

C. Work.

In SI units, energy is measured in Joules, i.e. (mass)(length squared)/(time squared). This is the same unit as is used for work. You can verify this by calculating that since work is force (which is mass times acceleration) times distance, the units work out to be the same. Force is measured in Newtons which are (mass)(distance)/(time squared); momentum is measured in (mass)(length)/(time); power is measured in Watts (which equal Joules/second or (mass)(distance squared)/time squared/time). Therefore, the **answer must be (C)**.

56. A ball rolls down a smooth hill. You may ignore air resistance. Which of the following is a true statement?
(Rigorous) (Skill 13.2)

A. The ball has more energy at the start of its descent than just before it hits the bottom of the hill, because it is higher up at the beginning.

B. The ball has less energy at the start of its descent than just before it hits the bottom of the hill, because it is moving more quickly at the end.

C. The ball has the same energy throughout its descent, because positional energy is converted to energy of motion.

D. The ball has the same energy throughout its descent, because a single object (such as a ball) cannot gain or lose energy.

C. The ball has the same energy throughout its descent, because positional energy is converted to energy of motion.

The principle of Conservation of Energy states that (except in cases of nuclear reaction, when energy may be created or destroyed by conversion to mass), "Energy is neither created nor destroyed, but may be transformed." Answers (A) and (B) give you a hint in this question—it is true that the ball has more Potential Energy when it is higher, and that it has more Kinetic Energy when it is moving quickly at the bottom of its descent. However, the total sum of all kinds of energy in the ball remains constant, if we neglect 'losses' to heat/friction.
Therefore, the **answer must be (C)**.

57. A long silver bar has a temperature of 50 degrees Celsius at one end and 0 degrees Celsius at the other end. The bar will reach thermal equilibrium (barring outside influence) by the process of heat _____.
(Rigorous) (Skill 13.2)

A. conduction.

B. convection.

C. radiation.

D. phase change.

A. conduction.

Heat conduction is the process of heat transfer via solid contact. The molecules in a warmer region vibrate more rapidly, jostling neighboring molecules and accelerating them. This is the dominant heat transfer process in a solid with no outside influences. Convection is heat transfer by way of fluid currents; radiation is heat transfer via electromagnetic waves; phase change can account for heat transfer in the form of shifts in matter phase. The answer to this question must **therefore be (A)**.

58. If the volume of a confined gas is increased, what happens to the pressure of the gas? You may assume that the gas behaves ideally, and that temperature and number of gas molecules remain constant.
(Easy) (Skill 13.4)

A. The pressure increases.

B. The pressure decreases.

C. The pressure stays the same.

D. There is not enough information given to answer this question.

B. The pressure decreases.

Because we are told that the gas behaves ideally, you may assume that it follows the Ideal Gas Law, i.e. $PV = nRT$. This means that an increase in volume must be associated with a decrease in pressure (i.e. higher T means lower P). Therefore, the **answer must be (B)**.

59. A Newton is fundamentally a measure of _____.
(Average Rigor) (Skill 13.5)

A. force.

B. momentum.

C. energy.

D. gravity.

A. Force.

In SI units, force is measured in Newtons. Momentum and energy each have different units, without equivalent dimensions. A Newton is one (kilogram)(meter)/(second squared), while momentum is measured in (kilgram)(meter)/(second) and energy, in Joules, is (kilogram)(meter squared)/(second squared). Although "gravity" can be interpreted as the force of gravity, i.e. measured in Newtons, fundamentally it is not required. Therefore, the **answer is (A)**.

60. Newton's Laws are taught in science classes because _____.
(Average Rigor) (Skill 14.1)

A. they are the correct analysis of inertia, gravity, and forces.

B. they are a close approximation to correct physics, for usual Earth conditions.

C. they accurately incorporate Relativity into studies of forces.

D. Newton was a well-respected scientist in his time.

B. They are a close approximation to correct physics, for usual Earth conditions.

Although Newton's Laws are often taught as fully correct for inertia, gravity, and forces, it is important to realize that Einstein's work (and that of others) has indicated that Newton's Laws are reliable only at speeds much lower than that of light At speeds close to the speed of light, Relativity considerations must be used. Therefore, the only correct **answer is (B)**.

61. Which of the following is most accurate?
(Rigorous) (Skill 14.1)

A. Mass is always constant; weight may vary by location.

B. Mass and weight are both always constant.

C. Weight is always constant; mass may vary by location.

D. Mass and weight may both vary by location.

A. Mass is always constant; weight may vary by location.

When considering situations exclusive of nuclear reactions, mass is constant (mass, the amount of matter in a system, is conserved). Weight, on the other hand, is the force of gravity on an object, which is subject to change due to changes in the gravitational field and/or the location of the object. Thus, the **best answer is (A)**.

62. All of the following are considered Newton's Laws *except* for:
(Average Rigor) (Skill 14.1)

A. An object in motion will continue in motion unless acted upon by an outside force.

B. For every action force, there is an equal and opposite reaction force.

C. Nature abhors a vacuum.

D. Mass can be considered the ratio of force to acceleration.

C. Nature abhors a vacuum.

Newton's Laws include his law of inertia (an object in motion (or at rest) will stay in motion (or at rest) until acted upon by an outside force) (A), his law that (Force)=(Mass)(Acceleration) (D), and his equal and opposite reaction force law (B). Therefore, the **answer to this question is (C)**, because "Nature abhors a vacuum" is not one of these.

63. All of the following measure energy *except* for _____
(Rigorous) (Skill 14.5)

A. joules.

B. calories.

C. watts.

D. ergs.

C. Watts.

Energy units must be dimensionally equivalent to (force)x(length), which equals (mass)x(length squared)/(time squared). Joules, Calories, and Ergs are all metric measures of energy. Watts, however, are units of power, i.e. Joules per Second. Therefore, the **answer is (C)**.

64. Sound can be transmitted in all of the following *except*
(Easy) (Skill 15.1)

A. air.

B. water.

C. a diamond.

D. a vacuum.

D. A vacuum.

Sound, a longitudinal wave, is transmitted by vibrations of molecules. Therefore, it can be transmitted through any gas, liquid, or solid. However, it cannot be transmitted through a vacuum, because there are no particles present to vibrate and bump into their adjacent particles to transmit the waves. This is consistent only with **answer (D)**.

65. Sound waves are produced by _____
(Easy) (Skill 15.2)

A. pitch.

B. noise.

C. vibrations.

D. sonar.

C. Vibrations.

Sound waves are produced by a vibrating body. The vibrating air molecules move back and forth parallel to the direction of motion of the wave as they pass the energy from adjacent air molecules closer to the source to air molecules farther away from the source. Therefore, the **answer is (C)**.

66. The speed of light is different in different materials. This is a result of

_____
(Average Rigor) (Skill 15.3)

A. interference.

B. refraction.

C. reflection.

D. relativity.

B. Refraction.

Refraction (B) is the bending of light because it hits a material at an angle which changes its speed. (This is analogous to a cart rolling on a smooth road. If it hits a rough patch at an angle, the wheel on the rough patch slows down first, leading to a change in direction.) Interference (A) is when light waves interfere with each other to form brighter or dimmer patterns; reflection (C) is when light bounces off a surface; relativity (D) is a general topic related to light speed and its implications, but not specifically indicated here. Therefore, the **answer is (B)**.

67. A converging lens produces a real image _____
(Rigorous) (Skill 15.4)

A. always.

B. never.

C. when the object is within one focal length of the lens.

D. when the object is further than one focal length from the lens.

D. When the object is further than one focal length from the lens.

A converging lens produces a real image whenever the object is far enough from the lens (outside one focal length) so that the rays of light from the object can hit the lens and be focused into a real image on the other side of the lens. When the object is closer than one focal length from the lens, rays of light do not converge on the other side; they diverge so a virtual image is formed where those diverging rays would have converged if they had originated behind the object. Thus, the correct **answer is (D)**.

68. The Doppler Effect is associated most closely with which property of waves?
(Rigorous) (Skill 15.5)

A. amplitude.

B. wavelength.

C. frequency.

D. intensity.

C. frequency.

The Doppler Effect accounts for an apparent increase in frequency when the distance between the wave source and wave receiver increases or an apparent decrease in frequency when the distance between a wave source and a wave receiver decreases. Meanwhile, the amplitude, wavelength, and intensity of the wave are not as relevant to this process (although moving closer to a wave source makes it seem more intense). The **answer to this question is (C)**.

69. Which of the following is *not* a necessary characteristic of living things?
(Average Rigor) (Skill 16.1)

A. Movement.

B. Reduction of local entropy.

C. Ability to cause local energy form changes.

D. Reproduction.

A. Movement.

There are many definitions of "life," but in all cases, a living organism reduces local entropy, changes chemical energy into other forms, and reproduces. Not all living things move, however, so the correct **answer is (A)**.

70. Resistance is measured in units called
(Easy) (Skill 16.2)

A. watts.

B. volts.

C. ohms.

D. current.

C. Ohms.

A watt is a unit of energy. Potential difference is measured in a unit called the volt. Current is the number of electrons per second that flow past a point in a circuit. An ohm is the unit for resistance. The correct **answer is (C)**.

71. Which is the correct order of methodology?
(Average Rigor) (Skill 17.1)

1. collecting data
2. planning a controlled experiment
3. drawing a conclusion
4. hypothesizing a result
5. re-visiting a hypothesis to answer a question

A. 1,2,3,4,5

B. 4,2,1,3,5

C. 4,5,1,3,2

D. 1,3,4,5,2

B. 4,2,1,3,5

The correct methodology for the scientific method is first to make a meaningful hypothesis (educated guess), then plan and execute a controlled experiment to test that hypothesis. Using the data collected in that experiment, the scientist then draws conclusions and attempts to answer the original question related to the hypothesis. This is consistent only with **answer (B)**.

72. Which of the following is *not* considered ethical behavior for a scientist?
(Average Rigor) (Skill 17.1)

A. Using unpublished data and citing the source.

B. Publishing data before other scientists have had a chance to replicate results.

C. Collaborating with other scientists from different laboratories.

D. Publishing work with an incomplete list of citations.

D. Publishing work with an incomplete list of citations.

One of the most important ethical principles for scientists is to cite all sources of data and analysis when publishing work. It is reasonable to use unpublished data (A), as long as the source is cited. Most science is published before other scientists replicate it (B), and frequently scientists collaborate with each other, in the same or different laboratories (C). Therefore, the **answer is (D)**.

73. Extensive use of antibacterial soap has been found to increase the virulence of certain infections in hospitals. Which of the following might be an explanation for this phenomenon?
(Average Rigor) (Skill 17.1)

A. Antibacterial soaps do not kill viruses.

B. Antibacterial soaps do not incorporate the same antibiotics used as medicine.

C. Antibacterial soaps kill a lot of bacteria, and only the hardiest ones survive to reproduce.

D. Antibacterial soaps can be very drying to the skin.

C. Antibacterial soaps kill a lot of bacteria, and only the hardiest ones survive to reproduce.

All of the answer choices in this question are true statements, but the question specifically asks for a cause of increased disease virulence in hospitals. This phenomenon is due to natural selection. The bacteria that can survive contact with antibacterial soap are the strongest ones, and without other bacteria competing for resources, they have more opportunity to flourish. Therefore, the **answer is (C)**.

74. When is a hypothesis formed? *(Easy) (Skill 17.2)*

A. Before the data is taken.

B. After the data is taken.

C. After the data is analyzed.

D. While the data is being graphed.

A. Before the data is taken.

A hypothesis is an educated guess, made before undertaking an experiment. The hypothesis is then evaluated based on the observed data. Therefore, the hypothesis must be formed before the data is taken, not during or after the experiment. This is consistent only with **answer (A).**

75. Which of the following is *not* an acceptable way for a student to acknowledge sources in a laboratory report?
(Easy) (Skill 17.2)

A. The student tells his/her teacher what sources s/he used to write the report.

B. The student uses footnotes in the text, with sources cited, but not in correct MLA format.

C. The student uses endnotes in the text, with sources cited, in correct MLA format.

D. The student attaches a separate bibliography, noting each use of sources.

A. The student tells his/her teacher what sources s/he used to write the report.

It may seem obvious, but students are often unaware that scientists need to cite all sources used. For the young adolescent, it is not always necessary to use official MLA format (though this should be taught at some point). Students may properly cite references in many ways, but these references must be in writing, with the original assignment. Therefore, the **answer is (A)**.

76. In a laboratory report, what is the abstract?
(Easy) (Skill 17.4)

A. The abstract is a summary of the report, and is the first section of the report.

B. The abstract is a summary of the report, and is the last section of the report.

C. The abstract is predictions for future experiments, and is the first section of the report.

D. The abstract is predictions for future experiments, and is the last section of the report.

A. The abstract is a summary of the report, and is the first section of the report.

In a laboratory report, the abstract is the section that summarizes the entire report. It appears at the very beginning of the report, even before the introduction, often on its own page (instead of a title page). This format is consistent with articles in scientific journals. Therefore, the **answer is (A).**

77. Identify the control in the following experiment: A student had four corn plants and was measuring photosynthetic rate (by measuring growth mass). Half of the plants were exposed to full (constant) sunlight, and the other half were kept in 50% (constant) sunlight.
(Average Rigor) (Skill 17.4)

A. The control is a set of plants grown in full (constant) sunlight.

B. The control is a set of plants grown in 50% (constant) sunlight.

C. The control is a set of plants grown in the dark.

D. The control is a set of plants grown in a mixture of natural levels of sunlight.

A. The control is a set of plants grown in full (constant) sunlight.

In this experiment, the goal was to measure how two different amounts of sunlight affected plant growth. The control in any experiment is the 'base case,' or the usual situation without a change in variable. Because the control must be studied alongside the variable, answers (C) and (D) are omitted (because they were not in the experiment). The **better answer is (A)**, because usually plants are assumed to have the best growth and their usual growing circumstances in full sunlight.

78. In an experiment measuring the growth of bacteria at different temperatures, what is the independent variable?
(Rigorous) (Skill 17.4)

A. Number of bacteria.

B. Growth rate of bacteria.

C. Temperature.

D. Light intensity.

C. Temperature.

The independent variable in an experiment is the entity that is changed by the scientist, in order to observe the effects (the dependent variable(s)). In this experiment, temperature is changed in order to measure growth of bacteria, so **(C) is the answer**. Note that answer (A) is the dependent variable, and neither (B) nor (D) is directly relevant to the question.

79. A scientific law _____
(Easy) (Skill 17.4)

A. proves scientific accuracy.

B. may never be broken.

C. may be revised in light of new data.

D. is the result of one excellent experiment.

C. may be revised in light of new data.

A scientific law is the same as a scientific theory, except that it has lasted for longer, and has been supported by more extensive data. Therefore, such a law may be revised in light of new data, and may be broken by that new data. Therefore, the **answer must be (C).**

80. Amino acids are carried to the ribosome in protein synthesis by:
(Rigorous) (Skill 18.1)

A. transfer RNA (tRNA).

B. messenger RNA (mRNA).

C. ribosomal RNA (rRNA).

D. transformation RNA (trRNA).

A. transfer RNA (tRNA).

The job of tRNA is to carry and position amino acids to/on the ribosomes. mRNA copies DNA code and brings it to the ribosomes; rRNA is in the ribosome itself. There is no such thing as trRNA. Thus, the **answer is (A)**.

81. Which is the most desirable tool to use to heat substances in a middle school laboratory?
(Average Rigor) (Skill 18.1)

A. Alcohol burner.

B. Freestanding gas burner.

C. Bunsen burner.

D. Hot plate.

D. Hot plate.

Due to safety considerations, the use of open flame should be minimized, so a hot plate is the best choice. The best **answer is (D)**.

82. Chemicals should be stored _____
(Easy) (Skill 18.1)

A. in the principal's office.

B. in a dark room.

C. in an off-site research facility.

D. according to their reactivity with other substances.

D. According to their reactivity with other substances.

Chemicals should be stored with other chemicals of similar properties (e.g. acids with other acids), to reduce the potential for either hazardous reactions in the store-room, or mistakes in reagent use. Certainly, chemicals should not be stored in anyone's office, and the light intensity of the room is not very important because light-sensitive chemicals are usually stored in dark containers. In fact, good lighting is desirable in a store-room, so that labels can be read easily. Chemicals may be stored off-site, but that makes their use inconvenient. Therefore, the best **answer is (D)**.

83. When measuring the volume of water in a graduated cylinder, where does one read the measurement?
(Average Rigor) (Skill 18.2)

A. At the highest point of the liquid.

B. At the bottom of the meniscus curve.

C. At the closest mark to the top of the liquid.

D. At the top of the plastic safety ring.

B. At the bottom of the meniscus curve.

To measure water in glass, your eye must be at the top surface, and ascertain the location of the bottom of the meniscus (the curved surface at the top of the water). The meniscus forms because water molecules adhere to the sides of the glass, which is a slightly stronger force than their cohesion to each other. This is consistent only with **answer (B).**

84. Who should be notified in the case of a serious chemical spill?
(Easy) (Skill 18.2)

A. The custodian.

B. The fire department or other municipal authority.

C. The science department chair.

D. The School Board.

B. The fire department or other municipal authority.

Although the custodian may help to clean up laboratory messes, and the science department chair should be involved in discussions of ways to avoid spills, a serious chemical spill may require action by the fire department or other trained emergency personnel. It is best to be safe by notifying them in case of a serious chemical accident. Therefore, the **best answer is (B)**.

85. In a science experiment, a student needs to dispense very small measured amounts of liquid into a well-mixed solution. Which of the following is the best choice for his/her equipment to use?
(Average Rigor) (Skill 18.2)

A. Buret with Buret Stand, Stir-plate, Stirring Rod, Beaker.

B. Buret with Buret Stand, Stir-plate, Beaker.

C. Volumetric Flask, Dropper, Graduated Cylinder, Stirring Rod.

D. Beaker, Graduated Cylinder, Stir-plate.

B. Buret with Buret Stand, Stir-plate, Beaker.

The most accurate and convenient way to dispense small measured amounts of liquid in the laboratory is with a buret, on a buret stand. To keep a solution well-mixed, a magnetic stir-plate is the most sensible choice, and the solution will usually be mixed in a beaker. Although other combinations of materials could be used for this experiment, **choice (B)** is thus the simplest and best.

86. A laboratory balance is most appropriately used to measure the mass of which of the following?
(Easy) (Skill 18.2)

A. Seven paper clips.

B. Three oranges.

C. Two hundred cells.

D. One student's elbow.

A. Seven paper clips.

Usually, laboratory/classroom balances can measure masses between approximately 0.01 gram and 1 kilogram. Therefore, answer (B) is too heavy and answer (C) is too light. Answer (D) is silly, but it is a reminder to instruct students not to lean on the balances or put their things near them. **Answer (A)**, which is likely to have a mass of a few grams, is correct in this case.

87. Who determines the laws regarding the use of safety glasses in the classroom? *(Average) (Skill 18.4)*

A. The state government.

B. The school site.

C. The federal government.

D. The local district.

A. The state government.

Health and safety regulations are set by the state government, and apply to all school districts **answer (A)**. Federal regulations may accompany specific federal grants, and local districts or school sites may enact local guidelines that are stricter than the state standards.

88. Formaldehyde should not be used in school laboratories for the following reason:
(Easy) (Skill 18.4)

A. it smells unpleasant.

B. it is a known carcinogen.

C. it is expensive to obtain.

D. it is an explosive.

B. It is a known carcinogen.

Formaldehyde is a known carcinogen, so it is too dangerous for use in schools. Although formaldehyde also smells unpleasant, a smell alone is not a definitive marker of danger. Furthermore, some odorless materials are toxic. Formaldehyde is neither particularly expensive nor explosive. Thus, the **answer is (B)**.

89. Experiments may be done with any of the following animals except
(Easy) (Skill 18.4)

A. birds.

B. invertebrates.

C .lower order life.

D. frogs.

A. Birds.

No dissections may be performed on living mammalian vertebrates or birds. Lower order life and invertebrates may be used. Biological experiments may be done with all animals except mammalian vertebrates or birds. Therefore the **answer is (A).**

90. Which of the following is the worst choice for a school laboratory activity?
(Average Rigor) (Skill 18.4)

A. A genetics experiment tracking the fur color of mice.

B. Dissection of a preserved fetal pig.

C. Measurement of goldfish respiration rate at different temperatures.

D. Pithing a frog to watch the circulatory system.

D. Pithing a frog to watch the circulatory system.

While any use of animals (alive or dead) must be done with care to respect ethics and laws, it is possible to perform choices (A), (B), or (C) with due care. However, modern practice precludes pithing animals (causing partial brain death while allowing some systems to function), as inhumane. Therefore, the answer to this **question is (D).**

91. In which situation would a science teacher be legally liable?
(Rigorous) (Skill 18.5)

A. The teacher leaves the classroom for a telephone call and a student slips and injures him/herself.

B. A student removes his/her goggles and gets acid in his/her eye.

C. A faulty gas line in the classroom causes a fire.

D. A student cuts him/herself with a dissection scalpel.

A. The teacher leaves the classroom for a telephone call and a student slips and injures him/herself.

Teachers are required to exercise a "reasonable duty of care" for their students. Accidents may happen (e.g. (D)), or students may make poor decisions (e.g. (B)), or facilities may break down (e.g. (C)). However, the teacher has the responsibility to be present and to do his/her best to create a safe and effective learning environment. Therefore, the **answer is (A)**.

92. Which of these is the best example of 'negligence'?
(Average Rigor) (Skill 18.5)

A. A teacher fails to give oral instructions to those with reading disabilities.

B. A teacher fails to exercise ordinary care to ensure safety in the classroom.

C. A teacher displays inability to supervise a large group of students.

D. A teacher reasonably anticipates that an event may occur, and plans accordingly.

B. A teacher fails to exercise ordinary care to ensure safety in the classroom.

'Negligence' is the failure to "exercise ordinary care" to ensure an appropriate and safe classroom environment. It is best for a teacher to meet all special requirements for disabled students, and to be good at supervising large groups. However, if a teacher can prove that s/he has done a reasonable job to ensure a safe and effective learning environment, then it is unlikely that she/he would be found negligent. Therefore, **the answer is (B)**.

93. Which item should always be used when handling glassware?
(Easy) (Skill 18.5)

A. Tongs.

B. Safety goggles.

C. Gloves.

D. Buret stand.

B. Safety goggles.

Safety goggles are the single most important piece of safety equipment in the laboratory, and should be used any time a scientist is using glassware, heat, or chemicals. Other equipment (e.g. tongs, gloves, or even a buret stand) has its place for various applications. Therefore, the **answer is (B)**.

94. Accepted procedures for preparing solutions involve the use of
(Average Rigor) (Skill 18.5)

A. alcohol.

B. hydrochloric acid.

C. distilled water.

D. tap water.

C. Distilled water.

Alcohol and hydrochloric acid should never be used to make solutions unless instructed to do so. All solutions should be made with distilled water as tap water contains dissolved particles which may affect the results of an experiment. The correct **answer is (C)**.

95. When designing a scientific experiment, a student considers all the factors that may influence the results. The process goal is to _____
(Easy) (Skill 19.2)

A. recognize and manipulate independent variables.

B. recognize and record independent variables.

C. recognize and manipulate dependent variables.

D. recognize and record dependent variables.

A. Recognize and manipulate independent variables.

When a student designs a scientific experiment, s/he must decide what to measure, and what independent variables will play a role in the experiment. S/he must determine how to manipulate these independent variables to refine his/her procedure and to prepare for meaningful observations. Although s/he will eventually record dependent variables (D), this does not take place during the experimental design phase. Although the student will likely recognize and record the independent variables (B), this is not the process goal, but a helpful step in manipulating the variables. It is unlikely that the student will manipulate dependent variables directly in his/her experiment (C), or the data would be suspect. Thus, the **answer is (A)**.

96. What is the scientific method?
(Easy) (Skill 19.3)

A. It is the process of doing an experiment and writing a laboratory report.

B. It is the process of using open inquiry and repeatable results to establish theories.

C. It is the process of reinforcing scientific principles by confirming results.

D. It is the process of recording data and observations.

B. It is the process of using open inquiry and repeatable results to establish theories.

Scientific research often includes elements from answers (A), (C), and (D), but the basic underlying principle of the scientific method is that people ask questions and do repeatable experiments to answer those questions and develop informed theories of why and how things happen. Therefore, the best **answer is (B)**.

97. Which of the following data sets is properly represented by a bar graph?
(Average Rigor) (Skill 19.3)

A. Number of people choosing to buy cars, vs. Color of car bought.

B. Number of people choosing to buy cars, vs. Age of car customer.

C. Number of people choosing to buy cars, vs. Distance from car lot to customer home.

D. Number of people choosing to buy cars, vs. Time since last car purchase.

A. Number of people choosing to buy cars, vs. Color of car bought.

A bar graph should be used only for data sets in which the independent variable is non-continuous (discrete), e.g. gender, color, etc. Any continuous independent variable (age, distance, time, etc.) should yield a scatter-plot when the dependent variable is plotted. Therefore, the **answer must be (A)**.

98. For her first project of the year, a student is designing a science experiment to test the effects of light and water on plant growth. You should recommend that she _____
(Average Rigor) (Skill 19.4)

A. manipulate the temperature also.

B. manipulate the water pH also.

C. determine the relationship between light and water unrelated to plant growth.

D. omit either water or light as a variable.

D. Omit either water or light as a variable.

As a science teacher for middle-school-aged kids, it is important to reinforce the idea of 'constant' vs. 'variable' in science experiments. Only one variable should be examined in each science experiment. Therefore it is counterproductive to add in other variables (answers (A) or (B)). It is also irrelevant to determine the light-water interactions aside from plant growth (C). So the only possible **answer is (D)**.

99. The theory of 'continental drift' is supported by which of the following?
(Rigorous) (Skill 20.3)

A. The way the shapes of South America and Europe fit together.

B. The way the shapes of Europe and Asia fit together.

C. The way the shapes of South America and Africa fit together.

D. The way the shapes of North America and Antarctica fit together.

C. **The way the shapes of South America and Africa fit together.**

The theory of 'continental drift' states that many years ago, there was one land mass on the earth ('pangea'). This land mass broke apart via earth crust motion, and the continents drifted apart as separate pieces. This is supported by the shapes of South America and Africa, which seem to fit together like puzzle pieces if you look at a globe. Note that answer choices (A), (B), and (D) give either land masses that do not fit together, or those that are still attached to each other. Therefore, the **answer must be (C)**.

XAMonline, INC. 21 Orient Ave. Melrose, MA 02176
Toll Free number 800-509-4128
TO ORDER Fax 781-662-9268 OR www.XAMonline.com

GEORGIA ASSESSMENTS FOR THE CERTIFICATION OF EDUCATORS   -GACE - 2008

PO#   Store/School:

Address 1:

Address 2 (Ship to other):

City, State Zip

Credit card number_____-_____-_____-_____   expiration_____
EMAIL _____
PHONE   **FAX**

13# ISBN 2007	TITLE	Qty	Retail	Total
978-1-58197-257-3	Basic Skills 200, 201, 202			
978-1-58197-528-4	Biology 026, 027			
978-1-58197-529-1	Science 024, 025			
978-1-58197-341-9	English 020, 021			
978-1-58197-569-7	Physics 030, 031			
978-1-58197-531-4	Art Education Sample Test 109, 110			
978-1-58197-545-1	History 034, 035			
978-1-58197-527-7	Health and Physical Education 115, 116			
978-1-58197-540-6	Chemistry 028, 029			
978-1-58197-534-5	Reading 117, 118			
978-1-58197-547-5	Media Specialist 101, 102			
978-1-58197-535-2	Middle Grades Reading 012			
978-1-58197-591-8	Middle Grades Science 014			
978-1-58197-345-7	Middle Grades Mathematics 013			
978-1-58197-686-1	Middle Grades Social Science 015			
978-158-197-598-7	Middle Grades Language Arts 011			
978-1-58197-346-4	Mathematics 022, 023			
978-1-58197-549-9	Political Science 032, 033			
978-1-58197-588-8	Paraprofessional Assessment 177			
978-1-58197-589-5	Professional Pedagogy Assessment 171, 172			
978-1-58197-259-7	Early Childhood Education 001, 002			
978-1-58197-587-1	School Counseling 103, 104			
978-1-58197-541-3	Spanish 141, 142			
978-1-58197-610-6	Special Education General Curriculum 081, 082			
978-1-58197-530-7	French Sample Test 143, 144			
			SUBTOTAL	
FOR PRODUCT PRICES GO TO WWW.XAMONLINE.COM			Ship	$8.25
			TOTAL	

CPSIA information can be obtained at www.ICGtesting.com
Printed in the USA
BVOW06s0912090315

390875BV00005B/137/P

9 781581 975840